ADVANCES IN CONTROL AND AUTOMATION OF WATER SYSTEMS

ADVANCES IN CONTROL AND AUTOMATION OF WATER SYSTEMS

Kaveh Hariri Asli, Faig Bakhman Ogli Naghiyev,
Reza Khodaparast Haghi, and Hossein Hariri Asli

Apple Academic Press

TORONTO   NEW JERSEY

© 2013 by
Apple Academic Press Inc.
3333 Mistwell Crescent
Oakville, ON L6L 0A2
Canada

Apple Academic Press Inc.
1613 Beaver Dam Road, Suite # 104
Point Pleasant, NJ 08742
USA

First issued in paperback 2021

Exclusive worldwide distribution by CRC Press, a Taylor & Francis Group

ISBN 13: 978-1-77463-209-3 (pbk)
ISBN 13: 978-1-926895-22-2 (hbk)

Library of Congress Control Number: 2012935664

Library and Archives Canada Cataloguing in Publication

Advances in control and automation of water systems/Kaveh Hariri Asli ... [et al.].

Includes bibliographical references and index.
ISBN 978-1-926895-22-2
1. Water-supply–Automation. 2. Water quality management–Automation. I. Asli, Kaveh Hariri

TD353.A38 2012   628.1   C2011-908699-9

Contents

List of Contributors

Kaveh Hariri Asli
National Academy of Science of Azerbaijan AMEA, Baku, Azerbaijan.

Hossein Hariri Asli
Applied Science University, Iran

Reza Khodaparast Haghi
University of Salford, United Kingdom

Faig Bakhman Ogli Naghiyev
Baku State University, Azerbaijan.

List of Abbreviations

EPS	Extended period simulation
FSI	Fluid-structure interpenetration
GIS	Geography information system
MOC	Method of characteristics
PLC	Programmable logic control
RTC	Real-time control
RWCT	Rigid water column theory
UFW	Unaccounted for water

Preface

This book provides a broad understanding of the main computational techniques used for processing Control and Automation of Water Systems. The theoretical background to a number of techniques is introduced and general data analysis techniques and examining the application of techniques in an industrial setting, including current practices and current research considered. The book also provides practical experience of commercially available systems and includes a small-scale water systems related projects.

The book offers scope for academics, researchers, and engineering professionals to present their research and development works that have potential for applications in several disciplines of hydraulic and mechanical engineering. Chapters ranged from new methods to novel applications of existing methods to gain understanding of the material and/or structural behavior of new and advanced systems.

This book will provide innovative chapters on the growth of educational, scientific, and industrial research activities among mechanical engineers and provides a medium for mutual communication between international academia and the water industry. This book publishes significant research reporting new methodologies and important applications in the fields of automation and control as well as includes the latest coverage of chemical databases and the development of new computational methods and efficient algorithms for hydraulic software and mechanical engineering.

1 A Numerical Exploration of Transient Decay Mechanisms in Water Distribution Systems

CONTENTS

NOMENCLATURES

λ = coefficient of combination,

t = time,

$\rho1$ = density of the light fluid (kg/m^3),

$\rho2$ = density of the heavy fluid (kg/m^3),

s = length,

τ = shear stress,

C = surge wave velocity (m/s),

$v2-v1$ = velocity difference (m/s),

e = pipe thickness (m),

K = module of elasticity of water(kg/m^2) ,

w = weight

$\lambda_{\circ}$ = unit of length

V = velocity

C = surge wave velocity in pipe

f = friction factor

$H2-H1$ = pressure difference (m-H_2O)

g = acceleration of gravity (m/s^2)

V = volume

Ee = module of elasticity(kg/m^2)

θ = mixed ness integral measure

C = wave velocity(m/s),
u = velocity (m/s),

D = diameter of each pipe (m),
θ = mixed ness integral measure,
R = pipe radius (m²),
J = junction point (m),
A = pipe cross-sectional area (m²)
d = pipe diameter(m),

Ev=bulk modulus of elasticity,
P = surge pressure (m),
C = velocity of surge wave (m/s),
ΔV= changes in velocity of water (m/s),
Tp = pipe thickness (m),
Ew = module of elasticity of water (kg/m²),
T = time (s),

σ = viscous stress tensor
c = speed of pressure wave (celerity-m/s)
f = Darcy–Weisbach friction factor
μ = fluid dynamic viscosity(kg/m.s)
γ= specific weight (N/m³)
I = moment of inertia (m^4)
r = pipe radius (m)
dp =is subjected to a static pressure rise
α = kinetic energy correction factor
ρ = density (kg/m³)
g=acceleration of gravity (m/s²)
K = wave number
Ep = pipe module of elasticity (kg/m²)
C1 = pipe support coefficient
ψ = depends on pipeline support-characteristics and Poisson's ratio

1.1   INTRODUCTION

Water hammer as fluid dynamics phenomena is an important case study for designer engineers. Water hammer is a disaster pressure surge or wave caused by the kinetic energy of a fluid in motion when it is forced to stop or change direction suddenly [1]. The majority of transients in water and wastewater systems are the result of changes at system boundaries, typically at the upstream and downstream ends of the system or at local high points. Consequently, results of present chapter can reduce the risk of system damage or failure with proper analysis to determine the system's default dynamic response. Design of protection equipment has helped to control transient energy. It has specified operational procedures to avoid transients. Analysis, design, and operational procedures all benefit from computer simulations in this chapter. The study of hydraulic transients is generally considered to have begun with the works of Joukowski (1898) [2] and Allievi (1902) [3]. The historical development of this subject makes for good reading. A number of pioneers have made breakthrough contributions to the field, including Angus, Parmakian (1963) [4] and Wood (1970) [5], who popularized and refined the graphical calculation method. Wylie and Streeter (1993) [6] combined the method of characteristics with computer modeling. The field of fluid transients is still rapidly evolving worldwide by Brunone et al. (2000) [7]; Koelle and Luvizotto, (1996) [8]; Filion and Karney, (2002) [9]; Hamam and McCorquodale, (1982) [10]; Savic and Walters, (1995) [11]; Walski and Lutes, (1994) [12]; Wu and Simpson, (2000) [13]. Various methods have been developed to solve transient flow in pipes. These ranges have been formed from approximate equations to numerical solutions of the nonlinear Navier–Stokes equations. In present chapter a computational approach is presented to analyze and record the transient flow (down to 5 milliseconds). Transient flow has been solved for pipeline in the range of approximate equations. These approximate equations have been solved by numerical solutions of the nonlinear Navier–Stokes equations in Method of Characteristics (MOC).

1.2 MATERIALS AND METHODS

The pilot subject in our study was: "Interpenetration of two fluids at parallel between plates and turbulent moving in pipe." For data collection process, Rasht city water main pipeline have been selected as Field Tests Model. Rasht city in the north of Iran was located in Guilan province (1,050,000 population). Data have been collected from the PLC of Rasht city water treatment plant. The pipeline was included water treatment plant pump station (in the start of water transmission line), 3.595 km of 2*1200 mm diameter pre-stressed pipes and one 50,000 m³ reservoir (at the end of water transmission line). All of these parts have been tied into existing water networks. Long-distance water transmission lines must be economical, reliable, and expandable. Therefore, present chapter shows safe hydraulic input to a network. This idea provides wide optimization and risk-reduction strategy for Rasht city main pipeline. Records were included multi-booster pressurized lines with surge protection ranging from check valves to gas vessels (one-way surge tank). This chapter has particular prospects for designing pressurized and pipeline segments. This means that by reduction of unaccounted for water (UFW), energy costs can be reduced. Experiences have been ensured reliable water transmission to the Rasht city main pipeline.

The MOC approach transforms the water hammer partial differential equations into the ordinary differential equations along the characteristic lines defined as the continuity equation and the momentum equation are needed to determine V and P in a one-dimensional flow system. Solving these two equations produces a theoretical result that usually corresponds quite closely to actual system measurements if the data and assumptions used to build the numerical model are valid. Transient analysis results that are not comparable with actual system measurements are generally caused by inappropriate system data (especially boundary conditions) and inappropriate assumptions. The MOC is based on a finite difference technique where pressures are computed along the pipe for each time step [14].

$$H_P = 1/2 \begin{pmatrix} C/g(V_{Le} - V_{ri}) + (H_{Le} + H_{ri}) \\ -C/g(f \, \Delta t/2D)(V_{Le} \, |V_{Le}| - V_{ri}|V_{ri}|) \end{pmatrix}, \tag{1}$$

$$V_P = 1/2 \begin{pmatrix} (V_{Le} + V_{ri}) + (g/c)(H_{Le} - H_{ri}) \\ -(f \, \Delta t/2D)(V_{Le} \, |V_{Le}| + V_{ri}|V_{ri}|) \end{pmatrix}, \tag{2}$$

Where
f=friction, C=slope (deg.), V=velocity, t=time, H=head (m).

1.2.1 Regression Equations

There is a relation between two or many Physical Units of variables. For example, there is a relation between volume of gases and their internal temperatures. The main approach in this research is investigation of relation between P—surge pressure (m) as a function "Y"— and several factors—as variables "X"—such as; ρ–density (kg.m^{-3}), C–velocity of surge wave (m.s^{-1}), g–acceleration of gravity (m.s^{-2}), ΔV–changes in velocity of water (m.s^{-1}), d–pipe diameter (m), T–pipe thickness (m), Ep–pipe module of elasticity (kg.m^{-2}), Ew–module of elasticity of water (kg.m^{-2}), C1–pipe support coefficient, T–Time (sec), Tp–pipe thickness (m). The investigation is needed when water hammer phenomena is happened.

In this study, fast transients, down to 5 milliseconds have been recorded. Methods such as, inverse transient calibration and leak detection in calculation of UFW has been used. Field tests have been formed on actual systems with flow and pressure data records. These comparisons require threshold and span calibration of all sensor groups, multiple simultaneous datum, and time base checks and careful test planning and interpretation. Lab model has recorded flow and pressure data (Table 1.1–1.2). The model is calibrated using one set of data and, without changing parameter values, it is used to match a different set of results [15].

Assumption (1): p=f (V), V–velocity (flow parameter) is the most important variable. Dependent Variable: P–pressure (bar), all input data are in relation to starting point of water hammer condition. The independent variable is Velocity (m/sec). Regression Software "SPSS 10.0.5" performs multidimensional scaling of proximity data to find least-squares representation of the objects in a low-dimensional space (Figure 1.1).

TABLE 1.1   Model Summary and Parameter Estimates (Start of water hammer condition).

Equation	Model Summary					Parameter Estimates			
	R Square	F	df1	df2	Sig.	a_0	a_1	a_2	a_3
Linear $y = a_0 + a_1 x$	.418	15.831	1	22	.001	6.062	.571		
Logarithmic(a)	.	.	.	.	.	.	.		
Inverse(b)	.	.	.	.	.	.	.		
Quadratic $y = a_0 + a_1 x + a_2 x^2$	.487	9.955	2	21	.001	6.216	-.365	.468	
Cubic $y = a_0 + a_1 x + a_2 x^2 + a_3 x^3$	.493	10.193	2	21	.001	6.239	.000	-.057	.174
Compound $A = Ce^{kt}$	.424	16.207	1	22	.001	6.076	1.089		
Power(a)	.	.	.	.	.	.	.		
S(b)	.	.	.	.	.	.	.		
Growth $(dA/dT) = KA$	.424	16.207	1	22	.001	1.804	.085		
Exponential $y = ab^x + g$	.424	16.207	1	22	.001	6.076	.085		
Logistic $y = ab^b + g$	.424	16.207	1	22	.001	.165	.918		

a– The independent variable contains non-positive values. The minimum value is .00. The Logarithmic and Power models cannot be calculated.

b– The independent variable contains values of zero. The Inverse and S models cannot be calculated. Regression Equation defined in stages (2-3-7-8) is meaningless. Stages (1-4-5-6-9-10-11) are accepted, because their coefficients are meaningful:

$$\text{Linear function} \therefore \text{pressure} = 6.062 + .571\text{Flow},$$
$$\text{Quadratic function} \therefore \text{pressure} = 6.216 - .365\text{Flow}^4 + .468\text{Flow}^3,$$
$$\text{Cubic function} \therefore \text{pressure} = 6.239 - .057\text{Flow}^2 + .174\text{Flow},$$
$$\text{Compound function} \therefore \text{pressure} = 1.089(1 + \text{Flow})^n, n = \text{compounding period}$$
$$\text{Growth function} \therefore \text{pressure} = 1.804(.085)^{\text{Flow}/.05},$$
$$\text{Exponential function} \therefore \text{pressure} = 6.076e^{\text{FlowLn.085}},$$
$$\text{Logitic function} \therefore \text{pressure} = 1/(1 + e^{-\text{Flow}}) \text{ or } \text{pressure} = .165 + .918\text{Flow}$$

(3)

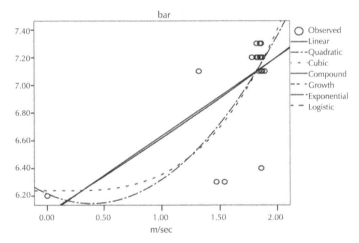

FIGURE 1.1   Scatter diagram for Tests for Water Transmission Lines (Field Tests Model).

Assumption (2): $p = f(V, T, L)$, V–velocity (flow) and T–time and L–distance, are the most important variables.

Input data are in relation with water hammer condition. Regression software fits the function curve (Figure 1.2–1.4) with regression analysis.

TABLE 1.2   Model Summary and Parameter Estimates (Water hammer condition).

Model		Un-standardized Coefficients	Standardized Coefficients		t	Sig.
			Std. Error	Beta		
1	(Constant)	28.762	29.73	-	0.967	0.346
	flow	0.031	0.01	0.399	2.944	0.009
	distance	-0.005	0.001	-0.588	-4.356	0
	time	0.731	0.464	0.117	1.574	0.133

TABLE 1.2 *(Continued)*

Model		Un-standardized Coefficients	Standardized Coefficients		t	Sig.
			Std. Error	Beta		
2	(Constant)	14.265	29.344	-	0.486	0.632
	flow	0.036	0.01	0.469	3.533	0.002
	distance	-0.004	0.001	-0.52	-3.918	0.001
3	(Constant)	97.523	1.519	-	64.189	0
4	(Constant)	117.759	2.114	-	55.697	0
	distance	-0.008	0.001	-0.913	-10.033	0
5	(Constant)	14.265	29.344	-	0.486	0.632
	flow	0.036	0.01	0.469	3.533	0.002
	distance	-0.004	0.001	-0.52	-3.918	0.001

Regression Equation defined in stage (1) is accepted, because its coefficients are meaningful:

$$Pressure = 28.762 + .031\ Flow - .005\ Distane + .731\ Time \tag{4}$$

TABLE 1.3 Model Summary and Parameter Estimates (Water hammer condition).

Pressure (m-Hd)	Flow (lit/sec)	Distance (m)	Time (sec)	Model	Variables Entered	Variables Removed	Method
				1	time, distance, flow(a)		Enter
86	2491	3390	0				
86	2491	3390	1				
88	2520	3291	0	2			Stepwise
90	2520	3190	1				(Criteria:
95	2574	3110	1.4				Probability-of-
95	2574	3110	1.4				F-to-enter <=
95	2574	3110	1.5			time	.050,
95	2590	3110	2				Probability-of-
95	2590	3110	2				F-to-remove
95.7	2600	3110	2				>= .100).
95.7	2600	3110	3	3	(a)	flow, distance(b)	Remove
95.7	2600	3110	4				
95.7	2600	3110	5	4			Forward
95.7	2605	3110	0.5				(Criterion:
100	2633	2184	1.3		distance		Probability-of-
100	2633	2928	1.3				F-to-enter <=
101	2650	2920	1.4				.050)
106	2680	1483	1.4				
107	2690	1217	1.4	5			Forward
109	2710	1096	1.4				(Criterion:
109	2710	1096	1.4		Flow		Probability-of-
110	2920	1000	1.5				F-to-enter <=
							.050)

a All requested variables entered.
b All requested variables removed.
c Dependent Variable: pressure

TABLE 1.4 Regression Model Summary and Parameter Estimates.

Model	R	R Square	Adjusted R Square	Std. Error of the Estimate
1	.955(a)	.912	.897	2.283
2	.949(b)	.900	.889	2.370
3	.000(c)	.000	.000	7.126
4	.913(d)	.834	.826	2.973
5	.949(b)	.900	.889	2.370

a Predictors: (Constant), time, distance, flow
b Predictors: (Constant), distance, flow
c Predictor: (Constant)
d Predictors: (Constant), distance

1.2.2 Regression

The Curve Estimation procedure allows quickly estimating regression statistics and producing related plots for 11 different models. Curve Estimation is most appropriate when the relationship between the dependent variable(s) and the independent variable is not necessarily linear.

TABLE 1.5 Regression Model Summary and Parameter Estimates (Water hammer condition).

Model		Sum of Squares	df	Mean Square	F	Sig.
1	Regression	972.648	3	324.216	62.223	.000(a)
	Residual	93.791	18	5.211		
	Total	1066.439	21			
2	Regression	959.744	2	479.872	85.455	.000(b)
	Residual	106.695	19	5.616		
	Total	1066.439	21			
3	Regression	.000	0	.000	.	.(c)
	Residual	1066.439	21	50.783		
	Total	1066.439	21			
4	Regression	889.663	1	889.663	100.655	.000(d)
	Residual	176.775	20	8.839		
	Total	1066.439	21			
5	Regression	959.744	2	479.872	85.455	.000(b)
	Residual	106.695	19	5.616		
	Total	1066.439	21			

a Predictors: (Constant), time, distance, flow
b Predictors: (Constant), distance, flow
c Predictor: (constant)
d Predictors: (Constant), distance
e Dependent Variable: pressure

- Linear regression is used to model the value of a dependent scale variable based on its linear relationship to one or more predictors.
- Nonlinear regression is appropriate when the relationship between the dependent and independent variables is not intrinsically linear.
- Binary logistic regression is most useful in modeling of the event probability for a categorical response variable with two outcomes.

The Auto regression procedure is an extension of ordinary least-squares regression analysis specifically designed for time series. One of the assumptions underlying ordinary least-squares regression is the absence of autocorrelation in the model residuals. Time series, however, often exhibit first-order autocorrelation of the residuals. In the presence of auto correlated residuals, the linear regression procedure gives inaccurate estimates of how much of the series variability is accounted for by the chosen predictors. This can adversely affect your choice of predictors and hence the validity of your model. The auto regression procedure accounts for first-order auto correlated residuals and provide reliable estimates of both goodness-of-fit measures and significance levels of chosen predictor variables.

TABLE 1.6 Regression Model Summary and Parameter Estimates Excluded Variables (Water hammer condition).

Model		Beta In	t	Sig.	Partial Correlation	Co- linearity Statistics
						Tolerance
2	time	.117(a)	1.574	.133	.348	.887
3	time	.122(b)	.552	.587	.122	1.000
	flow	.905(b)	9.517	.000	.905	1.000
	distance	-.913(b)	-10.033	.000	-.913	1.000
4	time	.189(c)	2.274	.035	.463	.995
	flow	.469(c)	3.533	.002	.630	.298
5	time	.117(a)	1.574	.133	.348	.887

a Predictors in the Model: (Constant), distance, flow
b Predictor: (constant)
c Predictors in the Model: (Constant), distance
d Dependent Variable: pressure

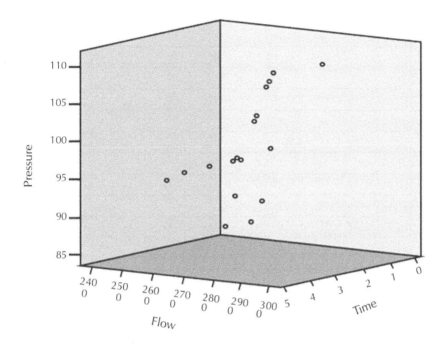

FIGURE 1.2   Scatter diagram for Lab. Tests (Research Field Tests Model).

1.2.3   Field Tests

Field tests can provide key modeling parameters such as the pressure-wave speed or pump inertia. Advanced flow and pressure sensors equipped with high-speed data loggers and PLC in water pipeline makes it possible to capture fast transients, down to 5 milliseconds. This research calibrated and validated numerical simulations for different fluids and systems for clients in the water and wastewater sectors. Comparisons between computer models and validation data can be grouped into the following three categories:

- Cases for which closed-form analytical solutions exist given certain assumptions. If the model can directly reproduce the solution, is considered valid for this case.
- Laboratory experiments compared with flow and pressure data records. The model is calibrated using one set of data and, without changing parameter values, it is used to match a different set of results. If successful, it is considered valid for these cases.
- Field tests on actual systems compared with flow and pressure data records. These comparisons require threshold and span calibration of all sensor groups, multiple simultaneous datum and time base checks and careful test planning and interpretation. Sound calibrations match multiple sensor records and reproduce both peak timing and secondary signals (all measured every second or fraction of a second).

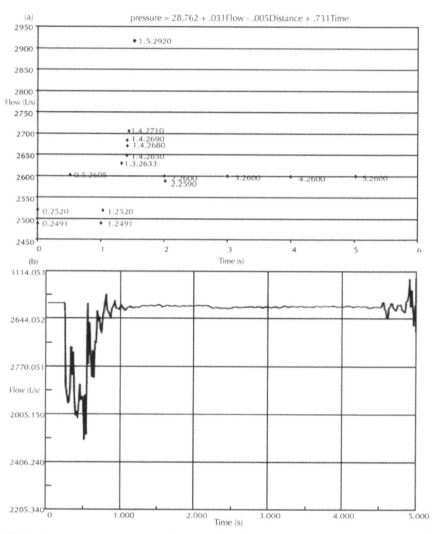

FIGURE 1.3 Analysis and Comparison results of calculations (a–Modeling results and b–Field Tests results, comparison) for Sangar–Saravan Water Pipeline pilot Research.

Field tests results showed water-column separation and the entrance of air into pipeline have been happened. Results showed that at point P25:J28 of Rasht city water-pipeline, air was interred to pipeline. Maximum volume of penetrated air was 198.483 (m³) and current flow was 2.666 (m³.s⁻¹). But in the second case, water-column separation did not happen. This was the effect of air release from the leakage location. It is noted that Max. Transient Pressure line was completely over the steady flow Pressure line. Maximum Pressure was 156.181(m). This was too high pressure for old piping and it must be consider as hazard for piping (Rasht city water-pipeline transmission line with surge tank and in leakage condition).

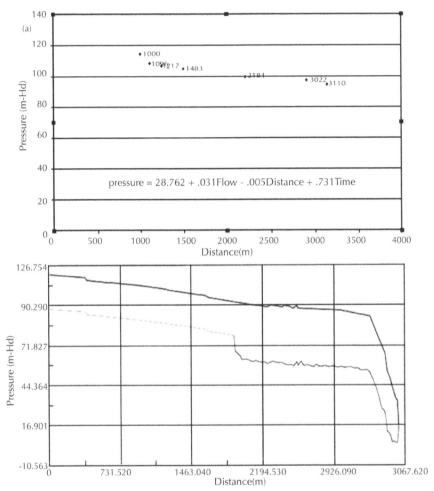

FIGURE 1.4 Rasht city Water Pipeline (a–Modeling results and b–Field Tests results, comparison).

Comparison between three parts of (Figure 1.5) for cases (Flow-Time, Head-Time, and Head-Distance transient curve) for Rasht city Water Pipeline proved the surge tank effective role. The First case was water-pipeline with water leakage and equipped with surge tank. The second case was water-pipeline without surge tank, but it had water leakage. The flow was decreased from 3014 (L.S^{-1}) down to Min. value 2520 (L.S^{-1}) after .6 (s). So in .4 (s), it was grown up to 3228 (L.S^{-1}). This was the effect of water release from the leakage location. Hence in one second, 494 (L.S^{-1}) water flows have been interred and exited to surge tank (for Transmission Line with surge tank and in leakage condition). It was shown that at 110M surge pressure in the near to pump station (start of Transmission Line), the leakage has been happened (Location of leakage). Hence, water flow was decreased from 3000 (L.S^{-1}) to 2500 (L.S^{-1}). This was UFW hazard alarm.

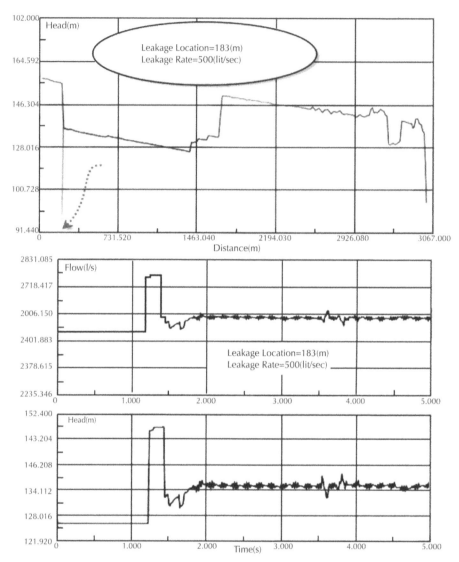

FIGURE 1.5 Rasht city Water Pipeline pilot Research (with surge tank and in leakage condition) Water flow decreased from 3000(l/s) to 2500(l/s).

In this chapter, location and rate of UFW in the pipeline was investigated. Location and rate of UFW in the pipeline are shown in Figure 1.5. Minimum Pressure line curve which was under the Transmission Line profile, located in vicinity of 50,000 m³ reservoir. Hence, there is a minus pressure in the zone of Transmission Line, in which it must be removed from the system. Maximum Transient Pressure line was completely over the steady flow pressure line. Maximum Pressure in the system was 156.181(m) .This pressure is too high for old pipes and it must be considered as a hazard for piping system. (Table 1.8–1.10).

1.2.4 LABORATORY MODELS

The model has been calibrated by water hammer Laboratory instrument. The model specifications are shown in (Table 1.7) and (Figure 1.8a).

TABLE 1.7 Laboratory Model Technical specifications.

Laboratory Model Technical specifications	Notation	Value	Dimension
pipe diameter	d	22	mm
surge tank cross section area	A	$1.521*10^{-3}$	m²
pipe cross section area	a	$.3204*10^{-3}$	m²
pipe thickness	t	0.9	mm
fluid density	ρ	1000	kg/ m³
volumetric coefficient	K	2.05	GN/ m²
fluid power	P	*	*
fluid force	F	*	*
friction loss	hf	*	*
frequency	W	*	*
fluid velocity	v	*	m/s
Max fluctuation	Ymax	*	*
flow rate	q	*	m³/s
pipe length	L	*	m
period of motion	T	*	*
Surge tank and reservoir elevation difference	y	*	m
surge wave velocity	C	*	m/s

* Laboratory experiments and Field Tests results

1.3 RESULTS AND DISCUSSION

The most important effects observed in Lab. and Field Tests is the influence of the rate of discharge from local leak to total discharge in the pipeline to the values of period of oscillations and wave celerity.

Research Field Tests Model (water pipeline of Rasht city in the north of Iran)

Software Hammer - Version 07.00.049.00

Type of Run: Full

Date of Run: 09/19/08

Time of Run: 04:47 am

Data File: E:\k-hariri Asli\ daraye nashti.inp

Hydrograph File: Not Selected

Labels: Short

TABLE 1.8 The paths in 26 points of Rasht city Water Pipeline, table Created by Hammer - Version 07.00.049.00.

No. of Pipe	From Points	Point	To Point	Length (m)			
P25	4	P25:J28	P25:N1	35.6			
| END |	MAX. PRESS. |	MIN. PRESS. |	MAX. HEAD |	MIN. HEAD |			
| POINT	| (mH) |	(mH) |	(m) |	(m) |			

P25:J28	42.4	0.0	137.6	95.2			
	** Air valve at node J28 **						
Time (s)	Volume (m3)	Head (m)	Mass (kg)	Air-Flow (cms)			
0.0000	.000	104.55	.0000	.000			
4.9885	.000	105.10	.0000	.000			
| FROM |	HEAD |	TO |	HEAD |	FLOW |	VEL		
| NODE |	(m) |	NODE |	(m) |	(cms) |	(m/s)|		
|	|	|	|	|	|	|	
| J28	| 112.8	| N1	| 112.6	| 2.500	| 2.21	|	

Point	Distance (m)	Elevation (m)	Init Head (m)	Max Head (m)	Min Head (m)	Max Vol3 (m3)	Vap Press (m)
+ P25:J28	.0	95.2	104.5	134.7	95.2	18.4	-10.0
P25:33.33%	11.9	95.5	104.5	134.4	93.8	.000	-10.0
P25:66.67%	23.7	95.7	104.4	121.1	95.0	.000	-10.0
P25:N1	35.6	95.9	104.3	104.3	104.3	.000	-

Note: Results showed at point P25:J28 of Rasht city Water Pipeline air was interred to pipeline. Maximum volume of Air was 198.483 (m³) and currently flow was 2.666 (m³.s⁻¹).

TABLE 1.9 Valves (at node J26-J9-J15-J17-J20-J28) data table Created by Hammer - Version 07.00.049.00 compared to equation of Regression software SPSS.

** Air valve at node J26 **				
Time (s)	Volume (m3)	Head (m)	Mass (kg)	Air-Flow (cms)
0.0000	.000	134.99	.0000	.000
4.9885	.000	134.95	.0000	.000
** Air valve at node J9 **				
Time (s)	Volume (m3)	Head (m)	Mass (kg)	Air-Flow (cms)
0.0000	.000	130.18	.0000	.000
4.9885	.000	129.75	.0000	.000

TABLE 1.9 *(Continued)*

		** Air valve at node J15 **		
Time	Volume	Head	Mass	Air-Flow
(s)	(m3)	(m)	(kg)	(cms)
0.0000	.000	115.22	.0000	.000
4.9885	.000	121.75	.0000	.000
		** Air valve at node J17 **		
Time	Volume	Head	Mass	Air-Flow
(s)	(m3)	(m)	(kg)	(cms)
0.0000	.000	113.01	.0000	.000
4.9885	.000	133.60	.0000	.000
		** Air valve at node J20 **		
Time	Volume	Head	Mass	Air-Flow
(s)	(m3)	(m)	(kg)	(cms)
0.0000	.000	112.65	.0000	.000
4.9885	.000	118.03	.0000	.000
		** Air valve at node J28 **		
Time	Volume	Head	Mass	Air-Flow
(s)	(m3)	(m)	(kg)	(cms)
0.0000	.000	104.55	.0000	.000
4.9885	.000	105.10	.0000	.000
		** Surge tank at node J4 **		
Time	Level	Head	Inflow	Spll-Rate
(s)	(m)	(m)	(cms)	(cms)
0.0000	135.0	135.0	.000	.000
4.9885	135.0	135.0	.002	.000

1.3.1 Influence of the Rate of Discharge from Local Leak on the Maximal Value of Pressure

The reason for the high deceasing in water Transmission pressure was related to the leakage condition in the Transmission Line (local leakage effect for high deceasing in water pressure in Rasht city Water Pipeline) [16].

TABLE 1.10 a) System Information Preference, b) Debug Information.

(a) ITEM	VALUE
Time Step (s)	
Automatic	0.0148

TABLE 1.10   *(Continued)*

(a) ITEM	VALUE
User-Selected n/a	
Total Number	339
Total Simulation Time (s)5.00	
Units	cms, m
Total Number of Nodes	27
Total Number of Pipes	26
Specific Gravity	1.00
Wave Speed (m/s)	1084
Vapor Pressure (m)	-10.0
Reports	
Number of Nodes	52
Number of Time Steps	All
Number of Paths	1
Output	
Standard	No
Cavities (Open/Close)	No
Adjustments	
Adjusted Variable	Length|
Warning Limit (%)	75.00
Calculate Transient Forces	No
Use Auxiliary Data File	Yes
(b) ITEM	VALUE
Tolerances	
Initial Flow Consistency Value	0.0006
Initial Head Consistency Value	0.030
Criterion for Fr. Coef. Flag	0.025
Adjustments	
Elevation Decrease 0.00	
Extreme Heads Display	
All Times	Yes
After First Extreme	No
Friction Coefficient	
Model	Steady
1,000,000 x Kinematic Viscosity	n/a
Debug Parameters	
Level	Null

TABLE 1.10 *(Continued)*

FROM NODE	HEAD (m)	TO NODE	HEAD (m)	FLOW (cms)	VEL (m/s)
J6	133.4	J7	133.0	2.500	2.21
J7	133.0	J8	131.2	2.500	2.21
J3	135.0	J4	135.0	3.000	2.65
J4	135.0	J26	135.0	3.000	2.65
J26	135.0	J27	133.5	2.500	2.21
J27	133.5	J6	133.4	2.500	2.21
J8	131.2	J9	131.1	2.500	2.21
J9	131.1	J10	129.2	2.500	2.21
J10	129.2	J11	128.0	2.500	2.21
J11	128.0	J12	126.0	2.500	2.21
J12	126.0	J13	124.1	2.500	2.21
J13	124.1	J14	123.8	2.500	2.21
J14	123.9	J15	120.4	2.500	2.21
J15	120.4	J16	120.2	2.500	2.21
J16	120.2	J17	118.9	2.500	2.21
J17	118.9	J18	118.8	2.500	2.21
J18	118.8	J19	118.7	2.500	2.21
J19	118.7	J20	118.6	2.500	2.21
J20	118.6	J21	115.7	2.500	2.21
J21	115.7	J22	114.3	2.500	2.21
J22	114.3	J23	113.4	2.500	2.21
J23	113.5	J24	113.3	2.500	2.21
J24	113.3	J28	112.8	2.500	2.21
J28	112.8	N1	112.6	2.500	2.21
J1	40.6	J2	40.6	3.000	2.65
J2	40.6	J3	40.6	3.000	2.65

1.3.2 Comparison of Present Research Results with Other Expert's Research

Comparison of present research results (water hammer software modeling and SPSS modeling), with other expert's research results, shows similarity and advantages:

1.3.2.1 *Arris S Tijsseling, Alan E Vardy, 2002*

Present research assumed three states in Field Tests; Transmission Line with surge tank, Water hammer in leakage, and no leakage condition. Present research results (Figure 1.6a) have been compared with the results reported by Arris S Tijsseling and Alan E Vardy, (2002). Comparison shows similarity in results [17].

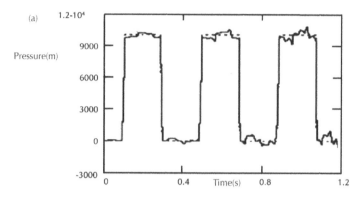

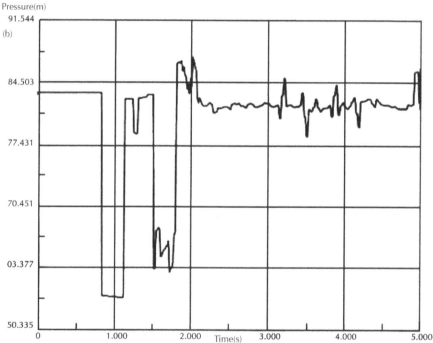

FIGURE 1.6 (a) pressure at midpoint; solid line = water hammer + 1D-FSI; dotted line = water hammer, (Arris S Tijsseling, Alan E Vardy Researches, 2002) (b) Transmission Line with surge tank and in leakage condition- water hammer happened condition present Research.

1.3.2.2   Arturo S. Leon, 2007

Comparison shows similarity between present chapter results and the results observed by Arturo S. Leon (2007) [18] (Figure 1.7a).

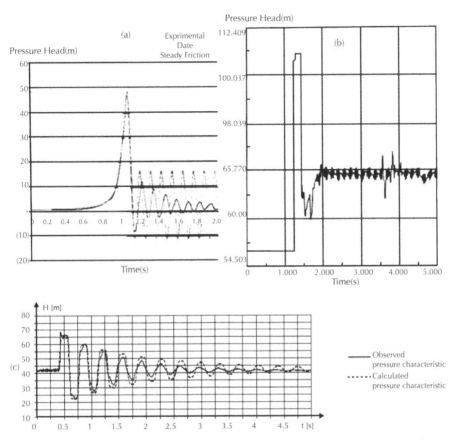

FIGURE 1.7 (a) Pressure Head Histories for a Single Piping System Using Steady and unsteady Friction. (Arturo S. Leon Research, 2007). (b) Rasht city. (c) pipeline with local leak. (Apoloniusz Kodura and Katarzyna Weinerowska, 2005)

1.3.2.3 Apoloniusz Kodura and Katarzyna Weinerowska, 2005

In present chapter, water hammer has been run in pressurized pipeline with the local leak. The experiments (Figure 1.8b) and numerical analysis results were presented. When pipeline with local leak was considered, the water hammer phenomenon have influenced by some additional factors [19]. Detailed conclusions were drawn on the basis of experiments and calculations for the pipeline with a local leak. Hence, the most important effects can be observed is that the influence of the ratio of discharge from local leak has been restricted. The outflow to the overpressure reservoir from the leak affects the value of wave celerity. (Figure 1.7c)

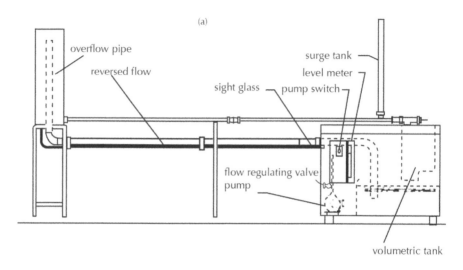

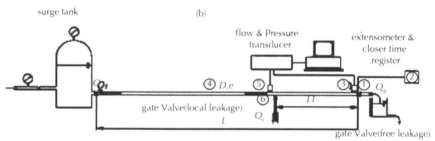

FIGURE 1.8 Scheme of the water hammer experimental equipment (a) Rasht city water hammer Laboratory Model, (b) water hammer Laboratory Model with local leak.(Kodura and Weinerowska)

1.3.2.4 Experimental Equipments and Conditions

Experiments were carried out in the laboratory of Warsa, University of Technology, Environmental Engineering Faculty, Institute of Water Supply and Water Engineering. (Figure 1.8b)

The physical model is schematically shown in (Figure 1.8b) .The main element was the pipeline (single straight pipe of the length L, extrinsic diameter D, and the wall thickness e or the pipeline consisted of sections of varied parameters. The pipeline was equipped with the valve at the end of the main pipe, which was joined with the closure time register. The water hammer pressure characteristics were measured by extensometers, and recorded in computer's memory. The supply of the water to the system was realized with use of reservoir which enabled inlet pressure stabilization. The experiments were carried out for four cases:

- simple positive water hammer for the straight pipeline of constant diameter; the measured characteristics were the basis for estimation the influence of the diameter change and local leak on water hammer run.

- positive water hammer in pipeline with single change of diameter: contraction and extension.
- positive water hammer in pipeline with local leak in two scenarios: with the outflow from the leak to the overpressure reservoir and with free outflow from the leak (to atmospheric pressure, with the possibility of sucking in air in negative phase). This was the reason for the sucking air in negative phase for Rasht city water-pipeline.

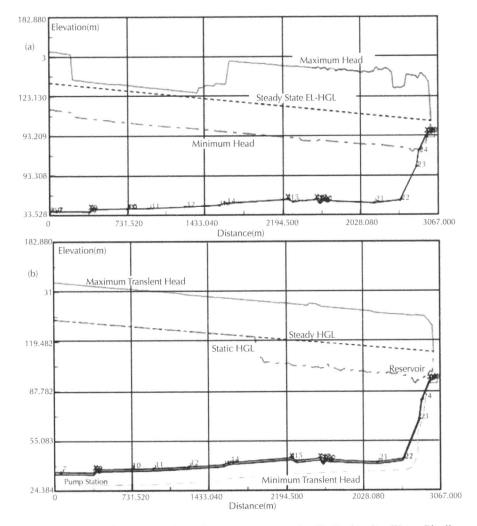

FIGURE 1.9 Column separations due to pump turned off—Rasht city Water Pipeline Transmission Line: (a) with surge tank—pipeline in leakage condition—pipeline was sucked air in negative phase, (b) without surge tank—pipeline in no leakage condition.

Column separations due to pump turned off for Rasht city Water Pipeline were carried out for two cases [20]:

(a) With surge tank and local leakage condition assumption: In this case air was sucked into the pipeline at negative phase. (Figure 1.9a)

(b) Without surge tank and local leakage condition assumption: In this case air was not sucked in the pipeline at negative phase. (Figure 1.9b)

1.4   CONCLUSION

It is always a good idea to run research to check extreme transient pressures. It is necessary for the system with large changes in elevation, long pipelines with large diameters (i.e., mass of water). Also, it is necessary for the initial (e.g., steady-state) velocities in excess of 1 (m/s). In some cases, hydraulic transient forces can result in cracks or breaks, even with low steady-state velocities.

In this chapter positive water hammer in pipeline was introduced with local leak in two scenarios. First, with the outflow from the leak to the overpressure reservoir, second: with free outflow from the leak (to atmospheric pressure, with the possibility of sucking air in negative phase). Field tests results show the leakage condition at negative phase air was sucked into the Rasht city water-pipeline (Figure 1.9a).There was minus pressure in a zone near the 50,000 m³ water reservoir (on point no. 124). So, this volume of air must be removed from the system.

KEYWORDS

- **Hazard alarm**
- **Kinetic energy**
- **Overpressure reservoir**
- **Transients**
- **Transmission line**

REFERENCES

1. Hariri, K. (2008). *Water hammer and fluid Interpenetration*. 9th Conference on Ministry of Energetic works at research week. Tehran, Iran.

2. Joukowski, N. (1898). Paper to Polytechnic Soc. Moscow.

3. Allievi, L. (1902). General theory of pressure variation in pipes.

4. Parmakian, J. (1963). *Water Hammer Analysis*. Dover Publications Inc., New York.

5. Wood, F. M. (1970). *History of Water Hammer*. Civil Engineering Research Report 65.

6. Wylie, E. B. and Streeter, V. L. (1993). Fluid Transients, Feb Press.

7. Brunone, B., Karney, B. W., Mecarelli, M., and Ferrante, M. (2000). Velocity profiles and unsteady pipe friction in transient flow. *Journal of Water Res. Plan. Mang. ASCE* **126**(4), 236-244.

8. Koelle, E., Luvizotto, E. Jr., and Andrade, J. P. G. (1996). Personality Investigation of Hydraulic Networks using MOC – Method of Characteristics. Proceedings of the 7th International Conference on Pressure Surges and Fluid Transients. Harrogate Durham, United Kingdom.

9. Filion, Y. and Karney, B. W. (2002). A numerical exploration of transient decay mechanisms in water distribution systems. Proceedings of the ASCE Environmental Water Resources Institute Conference. American Society of Civil Engineers. Roanoke, Virginia.

10. Hamam, M. A. and McCorquodale, J. A. (1982). Transient conditions in the transition from gravity to surcharged sewer flow.

11. Savic, D. A. and Walters, G. A. (1995). Genetic algorithms techniques for calibrating network models. Report No. 95/12, Centre for Systems and Control Engineering.

12. Walski, T. M. and Lutes, T. L. (1994). Hydraulic Transients Cause Low-Pressure Problems. *Journal of the American Water Works Association* 75(2), 58-62.

13. Wu, Z. Y. and Simpson, A. R. (2000). Evaluation of critical transient loading for optimal design of water distribution systems. Proceedings of the Hydro informatics conference. Iowa.

14. Joukowski, N. (1904). Paper to Polytechnic Soc. Moscow, spring of 1898, English translation by Miss O. Simin. Proc. AWWA.

15. Hariri, K. (2007). *Water hammer analysis and formulation.* 8th Conference on Ministry of Energetic works at research week. Tehran, Iran.

16. Wylie, E. B. and Streeter, V. L. (1982). Fluid Transients, Feb Press, Ann Arbor, MI, 1983. corrected copy: 166-171.

17. Arris S Tijsseling, "Alan E Vardy Time scales and FSI in unsteady liquid-filled pipe flow"5-12.

18. Arturo, S. Leon (2007) .An efficient second-order accurate shock-capturing scheme for modeling one and two-phase water hammer flows. Ph. D. Thesis, pp. 4-44.

19. Apoloniusz, Kodura, Katarzyna and Weinerowska, (2005). Some Aspects of Physical and Numerical Modeling of Water Hammer in Pipelines, pp. 126-132.

20. Kodura, A and Weinerowska, K. (2005). Some aspects of physical and numerical modeling of water hammer in pipelines. In: International symposium on water management and hydraulic engineering; pp. 125-33.

2 Mathematical Modeling of Hydraulic Transients in Simple Systems

CONTENTS

NOMENCLATURES

λ = coefficient of combination,

t = time,

$\rho 1$ = density of the light fluid (kg/m³),

$\rho 2$ = density of the heavy fluid (kg/m³),

s = length,

τ = shear stress,

C = surge wave velocity (m/s),

$v2-v1$ = velocity difference (m/s),

e = pipe thickness (m),

K = module of elasticity of water(kg/m²) ,

C = wave velocity(m/s),

u = velocity (m/s),

D = diameter of each pipe (m),

θ = mixed ness integral measure,

R = pipe radius (m²),

w = weight

λ_* = unit of length

V = velocity

C = surge wave velocity in pipe

f = friction factor

$H2-H1$ = pressure difference (m-H₂O)

g = acceleration of gravity (m/s²)

V = volume

Ee = module of elasticity(kg/m²)

θ = mixed ness integral measure

σ = viscous stress tensor

c = speed of pressure wave (celerity-m/s)

f = Darcy–Weisbach friction factor

μ = fluid dynamic viscosity(kg/m.s)

γ = specific weight (N/m³)

J = junction point (m),   I = moment of inertia (m^4)

A = pipe cross-sectional area (m²)   r = pipe radius (m)

d = pipe diameter(m), dp =is subjected to a static pressure rise

Ev = bulk modulus of elasticity,   α =kinetic energy correction factor

P = surge pressure (m),   ρ = density (kg/m³)

C = Velocity of surge wave (m/s),   g = acceleration of gravity (m/s²)

ΔV = changes in velocity of water (m/s),   K = wave number

Tp = pipe thickness (m),   Ep = pipe module of elasticity (kg/m²)

Ew = module of elasticity of water (kg/m²), C1=pipe support coefficient

T = time (s),   Ψ = depends on pipeline support-
                                           characteristics and Poisson's ratio

2.1 INTRODUCTION

Pump pulsation, due to interpenetration of two fluids, such as hydrodynamics insta-bility is a very loud banging, knocking, or hammering noise in the pipes that occurs when the flow is suddenly turned off. Pump pulsation is a sound like human heart by fluid pulsation in the pipe. This chapter refers to pump pulsation as a fluid transient in a "Dynamic" operating case, which may also include pulsation by fluid reciprocating or peristaltic positive displacement, sudden thrust due to relief valves that pop open or rapid piping accelerations due to an earthquake. It is advisable to investigate fluid-structure interpenetration (FSI) that can develop for dynamic cases but the decision to undertake such analysis is largely up to the designer; except for boilers or nuclear installations (Figure 2.1).

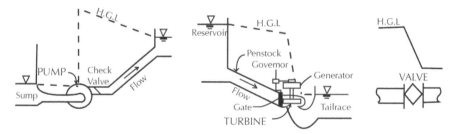

FIGURE 2.1   Common Cause of Hydraulic Transients.

Many researchers have studied water hammer and transient flow. Chaudhry and Hussaini (1985) have solved water hammer equations by MacCormak, Lambda, and Gabutti explicit Finite Difference (FD) schemes [1]. They found that these second-order FD schemes result in better solutions than the first-order. Izquierdo and Iglesias (2002) have developed a computer program to simulate hydraulic transients in a sim-ple pipeline system by mathematical modeling [2]. They also presented the users with a powerful tool to plan the potential risks to which an installation may be exposed and developed suitable protection strategies. Their model produced good numerical results within the accuracy of the used data.

This model was later generalized to include a pumping station fitted with check valve, delivery valve, and two air vessels. Izquierdo and Iglesias (2005), Ghidaoui, Mansour, and Zhao (2002) have proposed a two and five-layer eddy viscosity model for water hammer simulation [3, 4]. Zhao, Ghidaoui, and Godunov have formulated first- and second-order explicit finite volume (FV) methods of Godunov-type for water hammer problems [5, 6]. They compared the performances of FV schemes and Method of Charecteristics (MOC) schemes with space line interpolation for three test cases with and without friction. They modeled the wall friction using the formula of Brunone, Golia, and Greco (1991). The first-order FV Godunov-scheme produced the same results with MOC using space line interpolation. It was also shown that, for a given level of accuracy, the second-order Godunov-type scheme requires much less memory storage and execution time than the first-order Godunov-type scheme. Recently, Kodura and Weinerowska (2005) have investigated the difficulties that may arise in modeling of water hammer phenomenon [7]. Ghidaoui and Kolyshkin (2001) have performed linear stability analysis of the base flow velocity profiles for laminar and turbulent water hammer flows [8]. They found that the main parameters that govern the stability of the transient flows are the Reynolds number and a dimensionless time scale. This chapter provides a suitable way for detecting, analysis, and records of transient flow (down to 5 milliseconds) due to pump pulsation for water transmission line of Rasht city in the north of Iran. Transient flow has been solved for pipeline in the range of approximate equations. These approximate equations have been solved by numerical solutions of the nonlinear Navier–Stokes equations in the MOC.

2.2 MATERIALS AND METHODS

2.2.1 Field and Lab Tests for Disinfection of Water Transmission Lines

Field Tests: The Field Test was included water treatment plant pump station (in the start of water transmission line), 3.595 km of 2*1200mm diameter pre-stressed pipes and one 50,000 m³ reservoir (at the end of water transmission line). All of these parts have been tied into existing water networks.

Laboratory Model: A scale model have been built to reproduce transients observed in a prototype (real) system, typically for forensic or steam system investigations. This research Lab. model has recorded flow and pressure data. The model is calibrated using one set of data and, without changing parameter values. (Figure 2.2)

- **Laboratory Model Dateline:** The model has been calibrated and final checked by water hammer Laboratory Models.
- Sub-atmospheric leakage tests performed according to ASTM standards. This was done to explain repeated pipe breaks. This work led to improved standards for gasket designs and installation techniques in the province of sub atmospheric transient pressures which can suck contaminants into the water system [9].

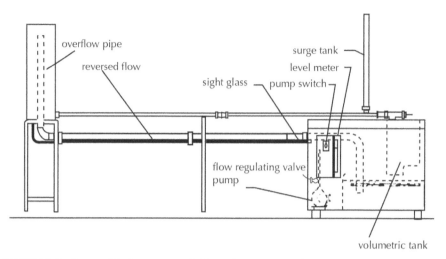

FIGURE 2.2   Research Laboratory Model for Laboratory Experiments with Flow and Pressure Data Records.

Laboratory Model Dateline: Dateline for Research Field Tests and Lab. model data collection was at 12:00 a.m., 10/02/07 until 05/02/09. Location of Research Field Tests and Lab. model was at Rasht city in the north of Iran. Pilot research subject was "Interpenetration of two fluids at parallel between plates and turbulent moving in pipe". For research data collection process, Rasht city water main pipeline have been selected as Research Field Tests Model. Rasht city in the north of Iran and located in Guilan province (1,050,000 population) [9]. Research data have been collected from the "PLC" of Rasht city water treatment plant.

The MOC is based on a FD technique where pressures are computed along the pipe for each time step,

$$Vp = \frac{1}{2}\left((V_{Le} + V_{ri}) + \left(\frac{g}{c}\right)(H_{Le} - H_{ri}) - \left(f\frac{\Delta t}{2D}\right)\left(V_{Le}|V_{Le}| + V_{ri}|V_{ri}|\right) \right), \qquad (21)$$

$$Hp = \frac{1}{2}\left(\frac{c}{g}(V_{Le} + V_{ri}) + (H_{Le} + H_{ri}) - \frac{c}{g}\left(f\frac{\Delta t}{2D}\right)\left(V_{Le}|V_{Le}| - V_{ri}|V_{ri}|\right) \right), \qquad (22)$$

Water hammer pressure or surge pressure (ΔH) is a function of independent variables ($\mathbf{X}$) such as:

$$\Delta H \approx f,\ T,\ C,\ V,\ g,\ D, \qquad (2.3)$$

For a model definition in this research relation between surge pressure (as a function) and several factors (as variables) have been investigated. Then Water hammer software has evaluated transient flow as a function of following parameters: ρ, K, d,

C1, Ee, V, f, T, C, g, Tp (Table 2.1–2.4). Regression software has fitted the function curve and provides regression analysis (Figure 2.3).

TABLE 2.1 Model Description of Regression software.

Model Name		MOD_2
Dependent Variable	1	bar
Equation	1	Linear
	2	Logarithmic
	3	Inverse
	4	Quadratic
	5	Cubic
	6	Compound(a)
	7	Power(a)
	8	S(a)
	9	Growth(a)
	10	Exponential(a)
	11	Logistic(a)
Independent Variable		m/sec
Constant		Included
Variable Whose Values Label Observations in Plots		Unspecified
Tolerance for Entering Terms in Equations		.0001

(a): The model requires all non-missing values to be positive.

2.2.2 Case Processing Summary

TABLE 2.2 Model Description of Regression software.

	N
Total Cases	24
Excluded Cases(a)	0
Forecasted Cases	0
Newly Created Cases	0

(a): Cases with a missing value in any variable are excluded from the analysis.

2.2.3   Variable Processing Summary

TABLE 2.3   Model Description of Regression software.

		Variables	
		Dependent	**Independent**
		bar	**m/sec**
Number of Positive Values		24	23
Number of Zeros		0	1(a,b)
Number of Negative Values		0	0
Number of Missing Values	User-Missing	0	0
	System-Missing	0	0

(a): The Inverse or S model cannot be calculated.

(b): The Logarithmic or Power model cannot be calculated.

2.2.4   Model Summary and Parameter Estimates

TABLE 2.4   Model Description of Regression software

Dependent Variable: bar									
Equation	**Model Summary**					**Parameter Estimates**			
	R Square	**F**	**df1**	**df2**	**Sig.**	a_0	a_1	a_2	a_3
Linear $y = a_0 + a_1 x$	.418	15.831	1	22	.001	6.062	.571		
Logarithmic(a)	.	.	.	.	.	.	.		
Inverse(b)	.	.	.	.	.	.	.		
Quadratic $y = a_0 + a_1 x + a_2 x^2$	.487	9.955	2	21	.001	6.216	-.365	.468	
Cubic $y = a_0 + a_1 x + a_2 x^2 + a_3 x^3$	.493	10.193	2	21	.001	6.239	.000	-.057	.174
Compound $A = Ce^{kt}$	.424	16.207	1	22	.001	6.076	1.089		
Power(a)	.	.	.	.	.	.	.		
S(b)	.	.	.	.	.	.	.		
Growth $(dA/dT) = KA$	.424	16.207	1	22	.001	1.804	.085		
Exponential $y = ab^x + g$	.424	16.207	1	22	.001	6.076	.085		
Logistic $y = ax^b + g$	.424	16.207	1	22	.001	.165	.918		

The independent variable is m/sec.

(a): The independent variable (m/sec) contains non-positive values. The minimum value is .00. The Logarithmic and Power models cannot be calculated.

(b): The independent variable (m/sec) contains values of zero. The Inverse and S models cannot be calculated.

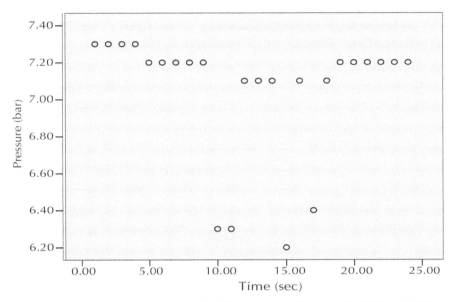

FIGURE 2.3   Regressions on Transmission Lines parameter (Research Field Tests Model)

2.3   RESULTS

Pulsation generally occurs when a liquid's motive force is generated by reciprocating or peristaltic positive displacement pumps. It is most commonly caused by the acceleration and deceleration of the pumped fluid. Installing a pulsation dampener can provide the most cost efficient and effective choice. It can prevent the damaging effects of pulsation. The most current pulsation dampener design is the hydro-pneumatic dampener, consisting of a pressure vessel containing a compressed gas, generally air or Nitrogen separated from the process liquid by a bladder or diaphragm.

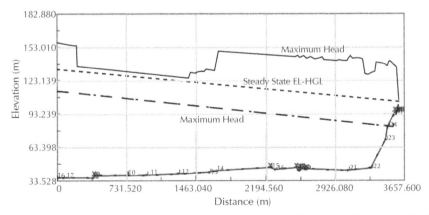

FIGURE 2.4   Column separations due to pump turned off. (Rasht city Water Pipeline Transmission Line with surge tank and in leakage condition)

The dampener can be installed as close as possible to the pump or quick closing valve and for charging to 85% of the liquid line pressure. Rapidly closing or opening a valve causes pressure transients in pipelines, known as water hammer.

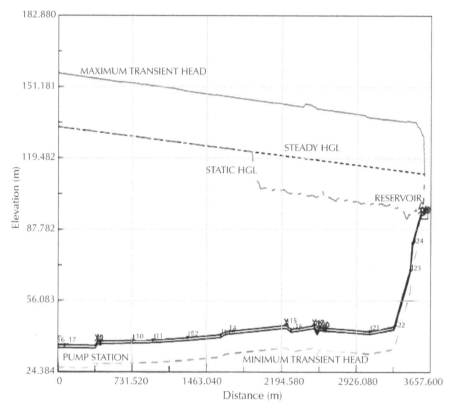

FIGURE 2.5   Rasht city Water Pipeline (Rasht city Water Pipeline Transmission Line without surge tank and in leakage condition)

Valve closure can result in pressures well over the steady state values, while valve opening can cause seriously low pressures, possibly so low that the flowing liquid vaporizes inside the pipe [10]. In this chapter, we have shown the maximum and minimum piezometric pressures, relative to atmospheric. This was observed in pipeline with respect to the time and location at which they were occurred. Results have been accounted helpful in design and determination of maximum (or minimum) expected pressures. Maximum or minimum points were due to the valve closure or opening and current pulsation [11]. Water hammer models of the Rasht city pipeline guided route selection, conceptual and detailed design of pipeline. Long-distance water transmission lines must be economical, reliable and expandable. These results were retained to provide hydraulic input to a network-wide optimization and risk-reduction strategy for Rasht city main pipeline. Data includes multi-booster pressurized lines with surge protection ranging from check valves to gas vessels (surge tank). Experimental results have been ensured reliable water transmission for the Rasht city main pipeline [12]. The main assumption was based on the exploration about transmission line witch was broken but equipped by pressure vessel or one way surge tank (real condition). Hence, water hammer has been analyzed in four manners:

1. water leakage assumption for transmission line,
2. No leakage assumption for transmission line,
3. water leakage assumption for transmission line witch was equipped by pressure vessel or one way surge tank (real condition).
4. No leakage for transmission line (this was equipped by pressure vessel or one way surge tank).

The initial condition and comparison between (Figure 2.4) and (Figure 2.5) are two cases of "Elevation-distance transient curve" for Rasht city. Water pipeline proved the surge tank effective role for Rasht city water pipeline. The first case was water pipeline with water leakage which was equipped with surge tank. The second case was water pipeline without surge tank and had water leakage.

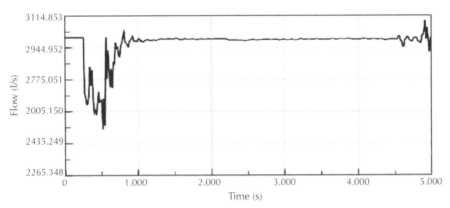

FIGURE 2.6 Field Tests Model: Flow-Time transient curve. (Rasht city Water Pipeline Transmission Line with surge tank and in leakage condition).

Results "in the second case" show water-column separation and the entrance of air into pipeline. It is observed that at point P25:J28 of Rasht city water-pipeline vapor was entered to pipeline. Maximum volume of Air was 198.483 (m³) and current flow was 2.666 (m³.s⁻¹) [13]. But "in the second case" water-column separation did not happen. This was due to the air release from the leakage location. Research showed Max. Transient Pressure line was completely over the steady flow Pressure line. Max. Pressure was 156.181(m). This pressure was too high for old piping and it must be considered as hazard for piping (Rasht city water-pipeline transmission line with surge tank and in leakage condition).

Comparison between (Figure 2.6) and (Figure 2.7) as two cases (Flow-Time transient curve) for Rasht city water-pipeline proved the surge tank effective role for Rasht city water-pipeline. The first case was water-pipeline with water leakage which was equipped with surge tank. The second case was water-pipeline without surge tank and had water leakage. The flow was decreased from 3,014 (l/s) down to minimum value 2,520 (l/s) after 6 (s). So in 4 (s), it was grown up to 3228 (l/s). This was the effect of water release from the leakage location. Hence in one second, 494 (l/s) water flows have been interred and exited to surge tank (for Transmission Line with surge tank and in leakage condition). It was showed that at 110M surge pressure in the near to pump station (start of Transmission Line) leakage has been happened (Location of leakage). Hence, water flow was decreased from 3,000 (l/s) to 2,500 (l/s). This was unaccounted for water (UFW) hazard.

This chapter is based on the two miscible liquids in acceleration and deceleration of the pumped fluid (Figure 2.8). So, interpenetration was investigated by comparison between theoretical hydraulics analysis and bench scale laboratory pilot [14-16].

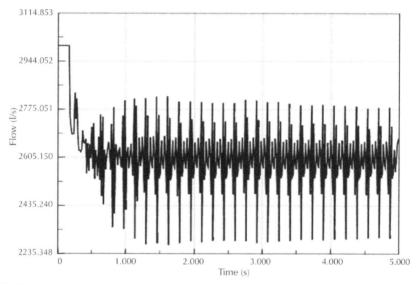

FIGURE 2.7   Field Tests Model: Flow-Time & Volume-Time transient curve. (Rasht city Water Pipeline Transmission Line without surge tank and in leakage condition).

Research Field Tests Model (water pipeline of Rasht city in the north of Iran)

Software   Hammer - Version 07.00.049.00

Type of Run: Full

Date of Run: 09/19/08

Time of Run: 04:47 am

Data File: E:\k-hariri Asli\ daraye nashti.inp

Hydrograph File: Not Selected

Labels: Short

TABLE 2.5   The paths in 26 points of Rasht city Water Pipeline, table Created by Hammer - Version 07.00.049.00.

No. of Pipe	Points	From Point	To Point	Length (m)
P2	5	P2:J6	P2:J7	60.7
P3	21	P3:J7	P3:J8	311.0
P4	2	P4:J3	P4:J4	1.0
P5	2	P5:J4	P5:J26	.5
P6	8	P6:J26	P6:J27	108.7
P7	3	P7:J27	P7:J6	21.5
P8	2	P8:J8	P8:J9	15.0
P9	23	P9:J9	P9:J10	340.7
P10	14	P10:J10	P10:J11	207.0
P11	22	P11:J11	P11:J12	339.0
P12	22	P12:J12	P12:J13	328.6
P13	4	P13:J13	P13:J14	47.0
P14	38	P14:J14	P14:J15	590.0
P15	4	P15:J15	P15:J16	49.0
P16	15	P16:J16	P16:J17	224.0
P17	2	P17:J17	P17:J18	18.4
P18	2	P18:J18	P18:J19	14.6
P19	2	P19:J19	P19:J20	12.0
P20	32	P20:J20	P20:J21	499.0
P21	17	P21:J21	P21:J22	243.4
P22	11	P22:J22	P22:J23	156.0
P23	3	P23:J23	P23:J24	22.0
P24	6	P24:J24	P24:J28	82.0
P25	4	P25:J28	P25:N1	35.6
P0	2	P0:J1	P0:J2	.5
P1	2	P1:J2	P1:J3	.5

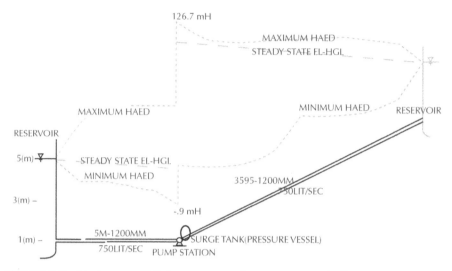

FIGURE 2.8   Maximum and Minimum pressure due to pump turned off.

TABLE 2.6   Output data table Created by Hammer - Version 07.00.049.00 Min. & Max. Head compared to equation of Regression software "SPSS".

END POINT	MAX. PRESS (mH)	MIN. PRESS (mH)	MAX. HEAD (m)	MIN. HEAD (m)
P2:J6	124.1	97.0	160.4	133.4
P2:J7	123.7	96.7	160.0	133.0
P3:J7	123.7	96.7	160.0	133.0
P3:J8	121.9	95.1	158.0	131.2
P4:J3	126.7	99.5	162.2	135.0
P4:J4	125.1	97.8	162.3	135.0
P5:J4	125.1	97.8	162.3	135.0
P5:J26	125.1	97.8	162.3	135.0
P6:J26	125.1	97.8	162.3	135.0
P6:J27	124.1	97.0	160.5	133.5
P7:J27	124.1	97.0	160.5	133.5
P7:J6	124.1	97.0	160.4	133.4
P8:J8	121.9	95.1	158.0	131.2
P8:J9	119.9	93.1	157.9	131.1
P9:J9	119.9	93.1	157.9	131.1
P9:J10	117.4	90.8	155.7	129.2
P10:J10	117.4	90.8	155.7	129.2
P10:J11	115.8	89.4	154.4	128.0
P11:J11	115.8	89.4	154.4	128.0
P11:J12	112.6	86.4	152.2	126.0
P12:J12	112.6	86.4	152.2	126.0
P12:J13	109.1	83.1	150.1	124.1

TABLE 2.6 *(Continued)*

END POINT	MAX. PRESS (mH)	MIN. PRESS (mH)	MAX. HEAD (m)	MIN. HEAD (m)
P13:J13	109.1	83.1	150.1	124.1
P13:J14	107.5	81.5	149.8	123.8
P14:J14	107.5	81.5	149.8	123.9
P14:J15	100.9	60.9	146.1	106.1
P15:J15	100.9	60.9	146.1	106.1
P15:J16	102.3	62.3	145.8	105.8
P16:J16	102.3	62.3	145.8	105.8
P16:J17	99.3	59.2	144.3	104.3
P17:J17	99.3	59.2	144.3	104.3
P17:J18	101.2	61.1	144.2	104.1
P18:J18	101.2	61.1	144.2	104.1
P18:J19	101.8	61.7	144.1	104.0
P19:J19	101.8	61.7	144.1	104.0
P19:J20	99.8	59.8	144.0	104.0
P20:J20	99.8	59.8	144.0	104.0
P20:J21	98.3	58.1	140.9	100.6
P21:J21	98.3	58.1	140.9	100.6
P21:J22	94.7	54.4	139.3	99.0
P22:J22	94.7	54.4	139.3	99.0
P22:J23	68.4	28.1	138.3	98.0
P23:J23	68.4	28.1	138.3	98.0
P23:J24	56.3	15.9	138.1	97.7
P24:J24	56.3	15.9	138.1	97.7
P24:J28	42.4	0.0	137.6	95.2
P25:J28	42.4	0.0	137.6	95.2
P25:N1	16.7	16.7	112.6	112.6
P0:J1	0.0	0.0	40.6	40.6
P0:J2	27.5	2.1	63.0	37.6
P1:J2	27.5	2.1	63.0	37.6
P1:J3	27.5	2.1	63.0	37.6

POINT	VAPOUR OR AIR	MAX. VOL* (m3)	CURR. VOL* (m3)	CURR. FLW (cms)
P25:J28	Air	198.483	.000	2.666

Maximum pressure in entire network is 126.7 mH at point P4:J3 of Rasht city Water Pipeline. Minimum pressure in entire network is –0.9 mH at point P25:J28 of Rasht city Water Pipeline. Elapsed time was 12 s for Hammer - Version 07.00.049.00 software. (Table 2.5–2.6) [17]

2.3.1   Comparison of Present research results with other expert's research

Comparison of present research results (water hammer software modeling) with other expert's research results shows similarity according to flowing:

Jaime Suárez Acuña and Chaudhry have obtained pressure heads by the steady and unsteady friction model. Comparison shows similarity in present Research and Jaime Suárez Acuña and Chaudhry (1987), research results [18] (Figure 2.9).

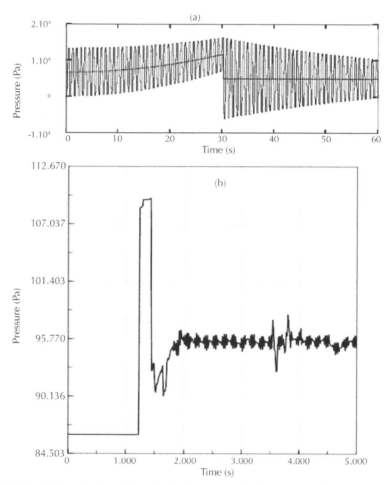

FIGURE 2.9   (a) Starting flow driven by 2 m/s velocity rise in 30 seconds time; average = rigid column; oscillation = water hammer at midpoint(Jaime Suárez Acuña and Chaudhry, 1987), (b) Transmission Line without surge tank and in leakage condition. (Present research -Rasht city Water Pipeline)

2.4 CONCLUSION

According to lengths and pressure-wave speeds of pipes in present research (Field Tests), it seems that it is necessary an advanced optimization algorithm. Algorithm considers lengths and pressure-wave speeds of pipes that must be installed at Rasht Water Transmission Line. Research suggests that advanced flow and pressure sensors with high-speed data loggers must be installed in Rasht Water Transmission Line. Data loggers must be linked to "PLC" in water pipeline. It will be recorded Pulsation and fast transients, down to 5 milliseconds for water flows interpenetration. Pressure transient recorder is a specialized data logger for monitoring rapid pressure changes in water pipe systems (i.e. Water Hammer) which is better to supplied in portable mode. Pressure Transient Loggers must be completely waterproof, submersible and battery powered .It must be maintenance free for at least five years. Pressure transient spikes are major cause of bursts and the associated expense of repair, water lost and interruptions to supply. Conventional loggers do not log rapidly enough to identify these transients which often last only seconds. For applications in Rasht Water Transmission Line Monitoring, MultiLog GPRSTM is ideal for monitoring flow and pressure. For water quality parameters monitoring to assess demand, leakage and conformance, MultiLog GPRSTM, can be used as well. For Network Analysis Investigations MultiLog GPRSTM can be used to perform dynamic flow and pressure analysis for Rasht Water Transmission Line models, particularly where hourly data updates can be useful for near real time monitoring.

KEYWORDS

- **Bladder**
- **Hydro-pneumatic dampener**
- **Pulsation**
- **Pumped fluid**
- **Thrust**

REFERENCES

1. Chaudhry, H. M. and Hussaini, M. Y. (1985). Second-order accurate explicit finite-difference schemes for water hammer analysis. *Journal of Fluids Engineering* 107, 523–529.
2. Izquierdo, J. and Iglesias, P. L. (2002). Mathematical modeling of hydraulic transients in simple systems. *Mathematical and Computer Modeling* 35(7), 801–812.
3. Ghidaoui, M. S., Mansour, G. S., and Zhao, M. (2002). Applicability of quasi steady and axisymmetric turbulence models in water hammer. *Journal of Hydraulic Engineering* 128(10), 917–924.
4. Izquierdo, J. and Iglesias, P. L. (2005). Mathematical modeling of hydraulic transients in complex systems. *Mathematical and Computer Modeling* 39(4), 529–540.
5. Zhao, M. and Ghidaoui, M. (2004). Godunov type solutions for water hammer flows. *Journal of Hydraulic Engineering, ASCE* 130(4), 341–348.

6. Brunone, B., Golia, U. M., and Greco, M. (1991). Some remarks on the momentum equation for fast transients. In: *Proceedings of International Conference on Hydraulic Transients with Water Column Separation, IAHR*. Valencia, Spain, pp. 201–209.

7. Kodura, A. and Weinerowska, K. (2005). Some aspects of physical and numerical modeling of water hammer in pipelines. In: *International symposium on water management and hydraulic engineering*, pp. 125–133.

8. Ghidaoui, M. S. and Kolyshkin, A. A. (2001). Stability analysis of velocity profiles in water hammer flows. *Journal of Hydraulic Engineering* 127(6), 499–512.

9. Hariri, K. (2007a). Decreasing of unaccounted for water "UFW" by Geographic Information System "GIS" in Rasht Urban Water System. Technical and Art. *J. Civil Engineering Organization of Gilan* 38, 3-7.

10. Hariri, K. (2007b). GIS and water hammer disaster at earthquake in Rasht water pipeline. 3rd International Conference on Integrated Natural Disaster Management (INDM). Tehran, Iran.

11. Hariri, K. (2007c). Interpenetration of two fluids at parallel between plates and turbulent moving in pipe. 8th Conference on Ministry of Energetic works at research week. Tehran, Iran.

12. Hariri, K. (2007d). Water hammer and valves. 8th Conference on Ministry of Energetic works at research week. Tehran, Iran.

13. Hariri, K. (2007e). Water hammer and hydrodynamics' instability. 8th Conference on Ministry of Energetic works at research week. Tehran, Iran.

14. Hariri, K. (2007f). Water hammer analysis and formulation. 8th Conference on Ministry of Energetic works at research week. Tehran, Iran.

15. Hariri, K. (2007g). Water hammer and fluid condition. 8th Conference on Ministry of Energetic works at research week. Tehran, Iran.

16. Hariri, K. (2007h). Water hammer and pump pulsation. 8th Conference on Ministry of Energetic works at research week. Tehran, Iran.

17. Hariri, K. (2007i). Reynolds number and hydrodynamics' instability. 8th Conference on Ministry of Energetic works at research week. Tehran, Iran.

18. Chaudhry, M. H. (1987). Applied Hydraulic Transients. Van Nostrand Reinhold, New York.

3 Modeling One and Two-phase Water Hammer Flows

CONTENTS

NOMENCLATURES

λ = coefficient of combination,

t = time,

$\rho 1$ = density of the light fluid (kg/m3),

$\rho 2$ = density of the heavy fluid (kg/m3),

s = length,

τ = shear stress,

C=surge wave velocity (m/s),

$v2-v1$=velocity difference (m/s),

e=pipe thickness (m),

K=module of elasticity of water(kg/m²),

C=wave velocity(m/s),

u = velocity (m/s),

D = diameter of each pipe (m),

θ = mixed ness integral measure,

R=pipe radius (m²),

w = weight

$\lambda_{\centerdot}$ = unit of length

V = velocity

C = surge wave velocity in pipe

f = friction factor

$H2-H1$=pressure difference (m-H2O)

g=acceleration of gravity (m/s²)

V=volume

Ee=module of elasticity(kg/m²)

θ = Mixed ness integral measure

σ = viscous stress tensor

c = speed of pressure wave (celerity- m/s)

f = Darcy–Weisbach friction factor

μ=fluid dynamic viscosity(kg/m.s)

γ= specific weight (N/m³)

J=junction point (m),   I= moment of inertia (m^4)

A = pipe cross-sectional area (m²)   r= pipe radius (m)

d=pipe diameter(m), dp =is subjected to a static pressure rise

Ev=bulk modulus of elasticity,   α =kinetic energy correction factor

P=surge pressure (m),   ρ= density (kg/m3)

C = velocity of surge wave (m/s),   g=acceleration of gravity (m/s²)

ΔV= changes in velocity of water (m/s),   K = wave number

Tp = pipe thickness (m),   Ep = pipe module of elasticity (kg/m2)

Ew = module of elasticity of water,   C1=pipe support coefficient

(kg/m2)

T=time (s),   Ψ= depends on pipeline support-
                                               characteristics and Poisson's ratio

q=flow rate (m³/s)

A=surge tank cross section area (m²)

y=surge tank and reservoir elevation difference (m)

a=pipe cross section area (m²)

L=pipe length (m)

hf=friction loss

W=frequency

T=period of motion

Ymax. =Max fluctuation

K=volumetric coefficient

P=fluid power

t=pipe thickness (mm)

F=fluid force

3.1   INTRODUCTION

The study of hydraulic transients is generally considered to have begun with the works of Joukowsky (1898) and Allievi (1902). The historical development of this subject makes for good reading. A number of pioneers made breakthrough contributions to the field, including R. Angus and John Parmakian (1963) and Wood (1970), who popularized and refined the graphical calculation method. Benjamin Wylie and Victor Streeter (1993) combined the method of characteristics with computer modeling. The field of fluid transients is still rapidly evolving worldwide by Brunone et al. (2000); Koelle and Luvizotto, (1996); Filion and Karney, (2002); Hamam and McCorquodale, (1982); Savic and Walters, (1995); Walski and Lutes, (1994); Wu and Simpson, (2000). Various methods have been developed to solve transient flow in pipes [1].

Governing Equations for Unsteady (or Transient) Flow—Hydraulic transient flow is also known as unsteady fluid flow. During a transient analysis, the fluid and system boundaries can be either elastic or inelastic:

- **Elastic theory** describes unsteady flow of a compressible liquid in an elastic system (e.g., where pipes can expand and contract). Research uses the Method of Characteristics (MOC) to solve virtually any hydraulic transient problems.

- **Rigid-column theory** describes unsteady flow of an incompressible liquid in a rigid system. It is only applicable to slower transient phenomena.

Both branches of transient theory stem from the same governing equations. The continuity equation and the momentum equation are needed to determine V and p in a one-dimensional flow system. Solving these two equations produces a theoretical result that usually corresponds quite closely to actual system measurements if the data and assumptions used to build the numerical model are valid. Transient analysis results that are not comparable with actual system measurements are generally caused by inappropriate system data (especially boundary conditions) and inappropriate assumptions.

Governing Equations for Steady-state Flow—Steady-state models, such as Water CAD or Water GEMS, are capable of two modes of analysis: steady-state and extended period simulation (EPS). The EPS solves a series of consecutive steady states using a gradient algorithm and accounting for mass in reservoirs and tanks (e.g., net inflows and storage). Both methods assume the system contains an incompressible fluid, so the total volumetric or mass inflows at any node must equal the outflows, less the change in storage. In addition to pressure head, elevation head, and velocity head, there may also be head added to the system, for instance, by a pump, and head removed from the system by friction. These changes in head are referred to as head gains and head losses, respectively. Balancing the energy across two points in the system yields the energy or Bernoulli equation for steady-state flow: the components of the energy equation can be combined to express two useful quantities, the hydraulic grade and the energy grade:

$$\left(P_1 / \gamma\right) + Z_1 \left(V_1^2 / 2g\right) + h_p = \left(P_2 / \gamma\right) + Z_2 + \left(V_2^2 / 2g\right) + h \qquad (3.1)$$

- **Hydraulic grade**—the hydraulic grade is the sum of the pressure head (p/γ) and elevation head (z). The hydraulic head represents the height to which a water column would rise in a piezometer. The plot of the hydraulic grade in a profile is often referred to as the hydraulic grade line or HGL.

- **Energy grade**—the energy grade is the sum of the hydraulic grade and the velocity head (V2/2g). This is the height to which a column of water would rise in a pitot tube. The plot of the hydraulic grade in a profile is often referred to as the energy grade line or EGL. At a lake or reservoir, where the velocity is essentially zero, the EGL is equal to the HGL [2, 3].

3.2   MATERIALS AND METHODS

3.2.1   Two Phase Flow

For Two phase flow systems with free gas and the potential for water-column separation, the numerical simulation of hydraulic transients is more complex and the computed results are more uncertain. Small pressure spikes caused by the type of tiny steam bubbles or vapor pockets that are difficult to simulate accurately seldom result in a significant change to the transient envelopes. Larger vapor-pocket collapse events resulting in significant upsurge pressures are simulated with enough accuracy to support definitive conclusions. The cause of a hydraulic transient is any sudden change in the fluid itself or any sudden change at the pressurized system's boundaries, including:

- Changes in fluid properties
- Changes at system boundaries

Hydraulic transients can result in the following physical phenomena: High or low transient pressures—These can be applied to piping and joints in a fraction of a second and they often alternate from high to low and *vice versa*. High pressures resulting from the collapse of vapor pockets are analogous to cavitations in a pump: they primarily accelerate wear and tear, but they can burst a pipe by overcoming its surge-tolerance limit. Sub atmospheric or even full-vacuum pressures can combine with overburden and groundwater pressures to collapse pipes by buckling failure. Groundwater can also be sucked into the piping.

High transient flows—these can result in significant degradation of water quality as deposits and rust are loosened and entrained at high velocities. This is aggravated whenever flows reverse direction during a transient event. High-velocity flows also exert rapidly moving pressure pulses result in temporary, but very significant, transient forces at bends and other fittings, which can cause joints to move. Even for buried pipe, repeated deflections combined with pressure cycling can wear out joints and result in leakage or outright failure. Thrust blocks are typically sized for steady-state forces plus a safety factor—not transient forces—and typically resist thrust in only one direction. In pump stations, low pressures on the downstream side of a slow-closing check valve may result in a very fast closure known as valve slam. A 10 psi (69 kPa) pressure differential across the face of a 16 in. (400 mm) valve can result in impact forces in excess of 2,000 lb. (8,900 N).

Column separation—Water columns typically separate at abrupt changes in profile or local high points due to sub atmospheric pressure. The space between the water columns is filled either by the formation of vapor (e.g., steam at ambient temperature) or air, if it is admitted to the pipeline through a valve. With vaporous forces at pipe bends Transient forces—cavitations, a vapor rocket forms and then collapses when the pipeline pressure increases as more flow enters the region than leaves it. Collapse of the vapor pocket can cause a dramatic high-pressure transient if the water column rejoins very rapidly, which can, in turn, cause the pipeline to rupture. Vaporous cavitations can also result in pipe flexure that damages pipe linings. High pressures can also result when air is expelled rapidly from a pipeline, which tends to repeat more times than when a vapor pocket collapses.

Vibrations—Rapid transient pressure fluctuations can result in vibrations or resonance that can cause even flanged pipes and fittings (bend and elbows) to dislodge, resulting in a leak or rupture. In fact, the cavitations that commonly occurs with water hammer can—as the phenomenon's name implies—release energy that sounds like someone pounding on the pipe with a hammer.

Rigid Column Theory—The rigid model assumes that the pipeline is not deformable and the liquid is incompressible. Therefore, system flow-control operations affect only the inertial and frictional aspects of transient flow. Given these considerations, it can be demonstrated using the continuity equation that any system flow-control operations results in instantaneous flow changes throughout the system, and that the liquid travels as a single mass inside the pipeline, causing a mass oscillation. If liquid density and pipe cross section are constant, the instantaneous velocity is the same in all sections. These rigidity assumptions result in an easy-to-solve ordinary differential equation; however, its application is limited to the analysis of surge. Newton's second law of motion is sufficient to determine the dynamic hydraulic of a rigid water body during the mass oscillation:

$$(L/D)(V|V|/2g)+(L/g)(dV/dt) \tag{3.2}$$

Elastic Theory—The elastic model assumes that changing the momentum of the liquid causes expansion or compression of the pipeline and liquid, both assumed to be linear-elastic. Since the liquid is not completely incompressible, its density can change slightly during the propagation of a transient pressure wave. The transient pressure wave will have a finite velocity that depends on the elasticity of the pipeline and of the liquid. In 1898, Joukowski established a theoretical relationship between pressure and velocity change during a transient flow condition.

$$H_2 - H_1 = (C/g)(V_2 - V_1) = \rho C(V_2 - V_1) \qquad \text{(3.3) (Joukowski Formula)}$$

In 1902, Allievi independently developed a similar elastic relation and applied it to a uniform valve closure.

$$C = 1/\left[\rho\left((1/k)+(dC_1/Ee)\right)\right]1/2 \qquad \text{(3.4) (Allievi Formula)}$$

The elastic theory developed by these two pioneers is fundamental to the field of hydraulic transients. The combined elasticity of both the water and the pipe walls is characterized by the pressure wave speed, a. This relation is a simplified form of the equation applicable to an instantaneous stoppage of velocity [4].

$$(H - H_0) = -a/g(V - V_0) \tag{3.5}$$

0=denotes initial conditions.

For an instantaneous valve closure or stoppage of flow, the upsurge pressure (H–H_0) is known as the "Joukowski head". Given that a is roughly 100 times as large as g, a 1 ft. Sec⁻¹ (0.3 m. Sec⁻¹) change in velocity can result in a 100 ft. (30 m) change in head. Because changes in velocity of several feet or meters per second when a pump shuts off or a hydrant or valve is closed, it is easy to see how large transients can occur readily in water systems. The mass of fluid that enters the part of the system located upstream of the valve immediately after its sudden closure is accommodated through the expansion of the pipeline due to its elasticity and through slight changes in fluid density due to its compressibility. This equation does not strictly apply to the drop in pressure downstream of the valve, if the valve discharges flow to the atmosphere.

Celerity and Pipe Elasticity—The elasticity of any medium is characterized by the deformation of the medium due to the application of a force. If the medium is a liquid, this force is a pressure force. The elasticity coefficient (also called the elasticity index, constant, or modulus) is a physical property of the medium that describes the relationship between force and deformation. Thus, if a given liquid mass in a given volume (V) is subjected to a static pressure rise (dp) a corresponding reduction ($dV <$ 0) in the fluid volume occurs. The relationship between cause (pressure increase) and effect (volume reduction) is expressed as the bulk modulus of elasticity (Ev) of the fluid, as given by:

$$E_v = -(dp/dV/V) = dp/d\rho \ / p \qquad (3.6)$$

A relationship between a liquid's modulus of elasticity and density yields its characteristic wave celerity:

$$a = \sqrt{Er/\rho} = \sqrt{dp/d\rho}, \qquad (3.7)$$

The characteristic wave celerity (a) is the speed with which a disturbance moves through a fluid. Its value is approximately 4,716 (ft.Sec^{-1}) or 1,438 (m. Sec^{-1}) for water and approximately 1,115 (ft.sec^{-1}) or 340 (m. Sec^{-1}) for air. Injecting a small percentage of small air bubbles can lower the effective wave speed of the fluid/air mixture, provided it remains well mixed. This is difficult to achieve in practice, because diffusers may malfunction and air bubbles may come out of suspension and coalesce or even buoy to the top of pipes and accumulate at elbows, for example. In 1848, Helmholtz demonstrated that wave celerity in a pipeline varies with the elasticity of the pipeline walls. Thirty years later, Korteweg developed an equation to determine wave celerity as a function of pipeline elasticity and liquid compressibility. Research uses an elastic model formulation that requires the wave celerity to be corrected to account for pipeline elasticity [5, 6].

$$a = \sqrt{E_0/\rho / \left(1 + E_r D\Psi \big/ E_e \right)} \qquad (3.8)$$

This is valid for thin walled pipelines ($D/e > 40$). The factor Ψ depends on pipeline support characteristics and Poisson's ratio. Ψ depends on the following:

- Pipe is anchored throughout against axial movement: $\Psi = 1 - \mu^2$, where μ is Poisson's ratio
- Pipe is equipped with functioning expansion joints throughout: $\Psi = 1 - \mu/2$

- Pipe is supported only at one end and allowed to undergo stress and strain both laterally and longitudinally:

$$\Psi = 5/4 - \mu \tag{3.9}$$

For thick-walled pipelines, various theoretical equations have been proposed to compute celerity; however, field investigations are needed to verify these equations. For pipes that exhibit significant visco-elastic effects (for example, plastics such as PVC and polyethylene), Covas et al. (2002) showed that these effects, including creep, can affect wave speed in pipes and must be accounted for if highly accurate results are desired. They proposed methods that account for such effects in both the continuity and momentum equations [7, 8].

3.2.2   Similar Work Presentations

- **Arithmetic method**—Assumes that flow stops instantaneously (in less than the characteristic time, 2 L/a), cannot handle water column separation directly, and neglects friction (Allievi, 1902; Joukowski, 1898).
- **Graphical method**—Neglects friction in its theoretical development but includes a means of accounting for it through a correction (Parmakian, 1963). It is time-consuming and not suited to solving networks or pipelines with complex profiles.
- **Design charts**—Provides basic design information for simple topologies at a few specific points (valve closure, pump and pipeline with no protection, surge tank, or air chamber protection). This method has been replaced by computer programs (Fok, 1978; Fok, 1980; Fok et al., 1982) based on the transient energy concept and backed by field and laboratory work (Fok, 1987).
- **Wave-plan method**—Represents initial transient disturbances as a series of pulses and tracks reflections at boundaries (Wood et al., 1966).
- **Method of Characteristics (MOC)**—most widely used and tested approach, with support for complex boundary conditions and friction and vaporous cavitations models. (PDEs) of continuity and momentum (e.g., Navier–Stokes) into ordinary differential equations that are solved algebraically along lines called characteristics. An MOC solution is exact along characteristics, but friction, vaporous cavitations and some boundary representations introduce errors in the results (Elansary, Silva, and Chaudhry, 1994; Gray, 1953; Streeter and Lai, 1962).
- **Field Tests**—Field tests can provide key modeling parameters such as the pressure-wave speed or pump inertia. Advanced flow and pressure sensors equipped with high-speed data loggers and "PLC" in water pipeline makes it possible to capture fast transients, down to 5 milliseconds. Methods such as inverse transient calibration and leak detection in calculation of Unaccounted for Water (UFW) use such data. Location for Field Tests and Lab. model was at Rasht city in the north of Iran. Pilot subject was named "Interpenetration of two fluids at parallel between plates and turbulent moving in pipe". For data collection process, Rasht city water main pipeline have been selected as Field Tests Model. Rasht city in the north of Iran was located in Guilan province (1,050,000

population). Dateline for Field Tests and Lab. model data collection have been started at 12:00 a.m., 10/02/07 until 05/02/09. Requirments data have been collected from the "PLC" of Rasht city water treatment plant.

Case Study—The pipeline was included water treatment plant pump station (in the start of water transmission line), 3.595 km of 2*1200 mm diameter pre-stressed pipes and one 50,000 m³ reservoir (in the end of water transmission line). All of these parts have been tied into existing water networks. This method provided a suitable way for detecting, analysis and records of transient flow (down to 5 milliseconds). Transient flow has been solved for pipeline in the range of approximate equations. These approximate equations have been solved by numerical solutions of the nonlinear Navier–Stokes equations in Method of Characteristics (MOC) [9]. The laboratory model specification data is shown in Table 3.1 and Figure 3.1.

TABLE 3.1   Laboratory Model Technical specifications.

Laboratory Model Technical specifications pipe diameter	Notation d	Value 22	Dimension mm
surge tank cross section area	A	$1.521*10^{-3}$	m²
pipe cross section area	a	$.3204*10^{-3}$	m²
pipe thickness	t	0.9	mm
fluid density	ρ	1000	kg/ m³
volumetric coefficient	K	2.05	GN/ m²
fluid power	P	*	*
fluid force	F	*	*
friction loss	hf	*	*
frequency	W	*	*
fluid velocity	v	*	m/s
Max fluctuation	Ymax	*	*
flow rate	q	*	m³/s
pipe length	L	*	m
period of motion	T	*	*
Surge tank and reservoir elevation difference	y	*	m
surge wave velocity	C	*	m/s

* Laboratory experiments and Field Tests results

If liquid density and pipe cross section are constant, the instantaneous velocity is the same in all sections. These rigidity assumptions result in an easy-to-solve ordinary differential equation; however, its application is limited to the analysis of surge. Newton's second law of motion is sufficient to determine the dynamic hydraulic of a rigid water body during the mass oscillation (Figure 3.2).

3.2.3 Laboratory Models

3.2.3.1 Newton Second Law for Laboratory Model

$$pal\frac{dv}{dt} = \rho gaH_1 - \rho ga(H_2 + y) + \rho gaL\sin\theta - \rho gah_f \tag{3.10.1}$$

If a steady-state flow condition is established—that is, if dV/dt = 0—then this equation simplifies to the Darcy-Weisbach formula for computation of head loss over the length of the pipeline. However, if a steady-state flow condition is not established because of flow control operations, then three unknowns need to be determined: H1 (t) (the left-hand head), H2 (t) (the right-hand head), V (t) (The instantaneous flow velocity in the conduit) to determine these unknowns, the engineer must know the boundary conditions at both ends of the pipeline. Using the fundamental rigid-model equation, the hydraulic grade line can be established for each instant. The slope of this line indicates the head loss between the two ends of the pipeline, which is also the head necessary to overcome frictional losses and inertial forces in the pipeline. For the case of flow reduction caused by a valve closure (dQ/dt < 0), the slope is reduced. If a valve is opened, the slope increases, potentially allowing vacuum conditions to occur. The change in slope is directly proportional to the flow change.

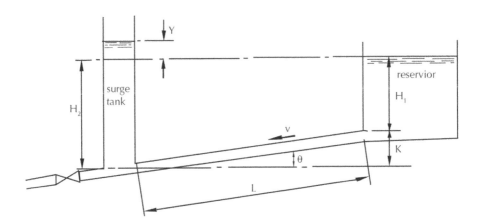

FIGURE 3.1 Laboratory Model for Laboratory experiments with flow and pressure data records

$$L\sin\theta = k, \frac{L}{g}\times\frac{dv}{dt} + y + h_f = 0, a.v = A\frac{dy}{dt} + Q,$$

$$\frac{L}{g} \times \frac{d}{dt}\left(\frac{A}{a}\frac{dy}{dt} + \frac{Q}{a}\right) + y + h_f = 0, \frac{d^2y}{dt^2} + \frac{ga}{LA}y = 0 \tag{3.10.2}$$

$$\frac{L}{g} \times \frac{d}{dt}\left(\frac{A}{a}\frac{dy}{dt} + \frac{Q}{a}\right) + y + h_f = 0, \frac{d^2y}{dt^2} + W^2 y = 0, \frac{d^2y}{dt^2} + \frac{ga}{LA}y = 0$$

$$\frac{d^2y}{dt^2} + \frac{ga}{LA}y = 0$$

$$\frac{d^2y}{dt^2} + W^2 y = 0, W^2 = \frac{ga}{LA}, \qquad \text{(If : f \& Q, Equal to zero)}$$

$$y = y_{max} \sin 2\pi \frac{t}{T} \tag{3.10.3},$$

$$T = \frac{2\pi}{W} = 2\pi\sqrt{\frac{LA}{ga}} \tag{3.10.4}$$

$$y_{max} = \frac{u}{W} = u\sqrt{\frac{LA}{ga}}, \quad y_{max} = y - 0.6h_f \tag{3.10.5},$$

Fluid power = $\qquad\qquad\qquad \delta.Q.h \qquad\qquad\qquad$ (3.10.6),

Fluid power = $\qquad\qquad\qquad \delta(H - \Delta h).V.A \qquad\qquad$ (3.10.7),

head friction = $\qquad\qquad\qquad f\frac{L}{D}\frac{V^2}{2g} \Rightarrow \qquad\qquad$ (3.10.8),

Fluid power = $\qquad\qquad\qquad \delta.V.A(H - KV^h) \qquad\qquad$ (3.10.9)

Generally, the maximum transient head envelope calculated by rigid water column theory (RWCT) is a straight line, as shown in the following figure. The rigid model has limited applications in hydraulic transient analysis because the resulting equations do not accurately model pressure waves caused by rapid flow control operations. The rigid model applies to slower surge or mass oscillation transients, as defined in "Wave Propagation and Characteristic Time".

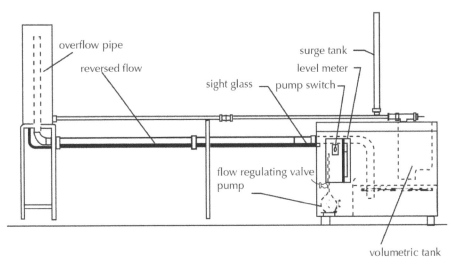

FIGURE 3.2   Laboratory Model for Laboratory experiments with flow and pressure data records

3.3   RESULTS AND DISCUSSION

Long-distance water transmission lines must be economical, reliable, and expandable. Therefore, a model was retained to provide hydraulic input to a network-wide optimization and risk-reduction strategy for Rasht city main pipeline. It track record includes multi-booster pressurized lines with surge protection ranging from check valves to gas vessels (surge tank). This chapter experiences have been ensured reliable water transmission for the Rasht city main pipeline. The water hammer phenomena investigation due to fluid condition was the main approach of this chapter. It formed by investigation of relation between: P-surge pressure (as a function or dependent variable with nomenclature "Y") and several factors (as independent variables with nomenclature "X") such as: ρ–density (kg/m³),C–velocity of surge wave (m/s), g–acceleration of gravity (m/s²), ΔV–changes in velocity of water (m/s), d–pipe diameter (m), T–pipe thickness (m), Ep–pipe module of elasticity (kg/m²), Ew–module of elasticity of water (kg/m²), C1–pipe support coefficient, T–Time (sec), Tp–pipe thickness (m) [10].

$$\text{Pressure} = 28.762 + .031 \text{ Flow} - .005 \text{ Distane} + .731 \text{ Time} \qquad (3.1.1)$$

Water Hammer Software Version 07.00.049.00 results have been compared with Regression software "SPSS". Regression software "SPSS" has fitted the function curve. It provided regression analysis. Field Tests have been compared by Lab. model results (Figure 3.3) [1].

3.3.1   Regression modeling results has been compared with Field Tests

Assumption: p = f (V, T, L), V–velocity (flow), T–time, and L–distance are the most important requested variables. Regression software fitted the function curve (Figure 3.4) and has provided regression analysis. Model have referred to a fluid transient as a "Dynamic" operating case, which may also include pulsation and sudden thrust due to relief valves that pop open or rapid piping accelerations (due to an earthquake). It is advisable to investigate fluid-structure interpenetration (FSI). This Model has been found the relation between two or many of variables accordance to fluid transient as a "Dynamic" operating. Results shown in Table 3.1 "Model Summary and Parameter Estimates" and (Table 3.2) [11].

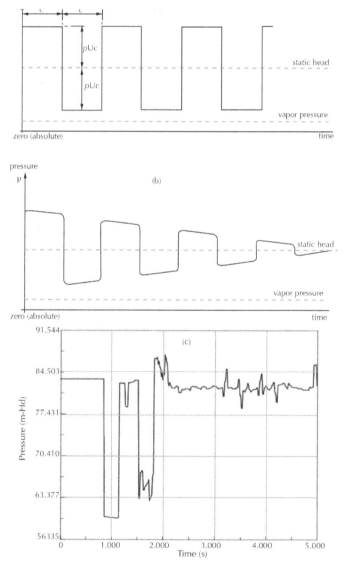

FIGURE 3.3   Laboratory Model surge wave in: (a) ideal system, (b) real system, (c) Rasht city water pipeline with surge tank and in leakage condition.

TABLE 3.2 Model Summary and Parameter Estimates (Water hammer condition).

Model		Un-standardized Coefficients	Standardized Coefficients		t	Sig.
			Std. Error	Beta		
1	(Constant)	28.762	29.73	-	0.967	0.346
	flow	0.031	0.01	0.399	2.944	0.009
	distance	−0.005	0.001	−0.588	−4.356	0
	time	0.731	0.464	0.117	1.574	0.133
2	(Constant)	14.265	29.344	–	0.486	0.632
	flow	0.036	0.01	0.469	3.533	0.002
	distance	−0.004	0.001	−0.52	−3.918	0.001
3	(Constant)	97.523	1.519	–	64.189	0
4	(Constant)	117.759	2.114	–	55.697	0
	distance	−0.008	0.001	−0.913	−10.033	0
5	(Constant)	14.265	29.344	–	0.486	0.632
	flow	0.036	0.01	0.469	3.533	0.002
	distance	−0.004	0.001	−0.52	−3.918	0.001

Regression Equation defined in stage (1) has been accepted, because its coefficients are meaningful:

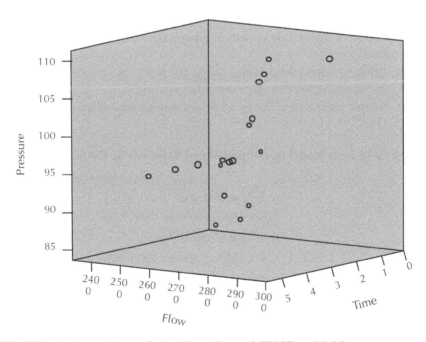

FIGURE 3.4 Scatter diagram for Lab Tests (Research Field Tests Model).

Hydraulic transient impacts can be expected at the following locations:

1. Check valves at pumps (as flow reverses from the downstream reservoir to the pump).
2. Reservoir inlet valves, altitude valves at elevated tanks, or isolation valves if they close rapidly.
3. Local high points where vapor or air pockets collapse.
4. Dead ends as they reflect incoming pulses with up to double the wave amplitude.
5. Pipe bursts, where flow leaving the system may exceed the steady-state flow (in systems with high static head compared to the dynamic head).
6. Surge-control devices if not properly designed or operated.
7. Changes in pipe-line profile or alignment (where transient forces may be significant).

Hydraulic transient impacts can be expected to occur at the following:

1. Pump startup before transient energy has decayed sufficiently or before all air has been removed from the line.
2. Pump emergency shutdown which may result in water-column separation and severe transient pressures due to vapor or air pocket formation and collapse.
3. Pump shifting during normal operations, which may result in frequent pressure shocks. Environmental concerns due to hydraulic transients include:

   a. Sewage spills or leaks to soils or groundwater during high transient pressures.
   b. Drinking water contamination due to air, debris, or groundwater intrusion during sub atmospheric pressures. (Table 3.3–3.5).
   c. Service interruptions due to repair and maintenance of infrastructure [12-13].

Research Field Tests Model (water pipeline of Rasht city in the north of Iran)

Software   Hammer - Version 07.00.049.00

Type of Run: Full

Date of Run: 09/19/08

Time of Run: 04:47 am

Data File: E:\k-hariri Asli\ daraye nashti.inp

Hydrograph File: Not Selected

Labels: Short

TABLE 3.3 (a) System Information Preference, (b) Debug Information.

(a)	
ITEM	VALUE
Time Step (s)	
Automatic	0.0148
User-Selected	n/a
Total Number	339
Total Simulation Time (s)	5.00
Units	cms, m
Total Number of Nodes	27
Total Number of Pipes	26
Specific Gravity	1.00
Wave Speed (m/s)	1084
Vapor Pressure (m)	−10.0
Reports	
Number of Nodes	52
Number of Time Steps	All
Number of Paths	1
Output	
Standard	No
Cavities (Open/Close)	No
Adjustments	
Adjusted Variable	Length
Warning Limit (%)	75.00
Calculate Transient Forces	No
Use Auxiliary Data File	Yes

(b)	
ITEM	VALUE
Tolerances	
Initial Flow Consistency Value	0.0006
Initial Head Consistency Value	0.030
Criterion for Fr. Coef. Flag	0.025
Adjustments	
Elevation Decrease	0.00
Extreme Heads Display	
All Times	Yes
After First Extreme	No
Friction Coefficient	
Model	Steady
1,000,000 x Kinematic Viscosity	| n/a
Debug Parameters	
Level	Null

TABLE 3.4 Initial Pipe Conditions Information Created by Hammer–Version 07.00.049.00 (Snapshot for selected end points at time 4.9885 s) compared to equation of Regression software "SPSS".

FROM NODE	HEAD (m)	TO NODE	HEAD (m)	FLOW (cms)	VEL (m/s)
J6	133.4	J7	133.0	2.500	2.21
J7	133.0	J8	131.2	2.500	2.21
J3	135.0	J4	135.0	3.000	2.65
J4	135.0	J26	135.0	3.000	2.65
J26	135.0	J27	133.5	2.500	2.21
J27	133.5	J6	133.4	2.500	2.21
J8	131.2	J9	131.1	2.500	2.21
J9	131.1	J10	129.2	2.500	2.21
J10	129.2	J11	128.0	2.500	2.21
J11	128.0	J12	126.0	2.500	2.21
J12	126.0	J13	124.1	2.500	2.21
J13	124.1	J14	123.8	2.500	2.21
J14	123.9	J15	120.4	2.500	2.21
J15	120.4	J16	120.2	2.500	2.21
J16	120.2	J17	118.9	2.500	2.21
J17	118.9	J18	118.8	2.500	2.21
J18	118.8	J19	118.7	2.500	2.21
J19	118.7	J20	118.6	2.500	2.21
J20	118.6	J21	115.7	2.500	2.21
J21	115.7	J22	114.3	2.500	2.21
J22	114.3	J23	113.4	2.500	2.21
J23	113.5	J24	113.3	2.500	2.21
J24	113.3	J28	112.8	2.500	2.21
J28	112.8	N1	112.6	2.500	2.21
J1	40.6	J2	40.6	3.000	2.65
J2	40.6	J3	40.6	3.000	2.65

TABLE 3.5 Valves (at node J26-J9-J15-J17-J20-J28) data table Created by Hammer–Version 07.00.049.00 compared to equation of Regression software "SPSS".

** Air valve at node J26 **				
Time (s)	Volume (m3)	Head (m)	Mass (kg)	Air-Flow (cms)
0.0000	.000	134.99	.0000	.000
4.9885	.000	134.95	.0000	.000

TABLE 3.5 *(Continued)*

		** Air valve at node J9 **		
Time	Volume	Head	Mass	Air-Flow
(s)	(m3)	(m)	(kg)	(cms)
0.0000	.000	130.18	.0000	.000
4.9885	.000	129.75	.0000	.000
		** Air valve at node J15 **		
Time	Volume	Head	Mass	Air-Flow
(s)	(m3)	(m)	(kg)	(cms)
0.0000	.000	115.22	.0000	.000
4.9885	.000	121.75	.0000	.000
		** Air valve at node J17 **		
Time	Volume	Head	Mass	Air-Flow
(s)	(m3)	(m)	(kg)	(cms)
0.0000	.000	113.01	.0000	.000
4.9885	.000	133.60	.0000	.000
		** Air valve at node J20 **		
Time	Volume	Head	Mass	Air-Flow
(s)	(m3)	(m)	(kg)	(cms)
0.0000	.000	112.65	.0000	.000
4.9885	.000	118.03	.0000	.000
		** Air valve at node J28 **		
Time	Volume	Head	Mass	Air-Flow
(s)	(m3)	(m)	(kg)	(cms)
0.0000	.000	104.55	.0000	.000
4.9885	.000	105.10	.0000	.000
		** Surge tank at node J4 **		
Time	Level	Head	Inflow	Spll-Rate
(s)	(m)	(m)	(cms)	(cms)
0.0000	135.0	135.0	.000	.000
4.9885	135.0	135.0	.002	.000

3.3.2 Comparison of Present research results with other expert's research

Comparison of present research results (water hammer software modeling and SPSS modeling), with other expert's research results, shows similarity and advantages (Figure 3.5) [13].

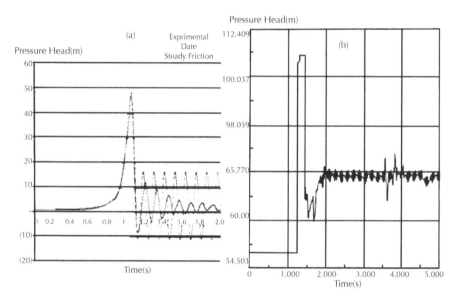

FIGURE 3.5   (a) Pressure Head Histories for a Single Pipe system, Using Steady and unsteady Friction. (Arturo S. Leon Research, 2007). (b) Rasht city Water Pipeline.

3.4   CONCLUSION

Hydraulic transients can result in the following infrastructure management issues and risks:

1. Premature aging and wear of valves, pipes, and pumps due to high magnitude and/ or frequent pressure shocks.

2. Pump cavitations due to low suction head and pipe lining damage due to vacuum conditions.

3. Rapid pump or valve operation by major water users (e.g., a food production factory) may accelerate the pipe material and anchor fatigue in their vicinity.

Rapidly closing or opening a valve causes pressure transients in pipelines, known as water hammer. Valve closure can result in pressures well over the steady state values, while valve opening can cause seriously low pressures, possibly so low that the flowing liquid vaporizes inside the pipe. Results have been showed the maximum and minimum piezometric pressure (relative to atmospheric). Hydrodynamics instabilities in a pipeline were occurred. Flow-Time transient curve (for Transmission Line with surge tank and in leakage condition) has been showed surge pressure equal to 110 (M-H$_2$O). Hence, water flow decreased from 3000 (l/s) to 2500 (l/s). This means UFW hazard was a serious alarm. The valuable observation was the action of leakage point as an open valve near the water treatment plant pump station. This open valve has been helped to release of air at exceeding of water pressure. But when vapor pressure was generated, air has been sucked into the pipeline.

KEYWORDS

- **Mass oscillation**
- **Piezometric pressure**
- **Pitot tube**
- **Unsteady flow**
- **Visco-elastic effects**

REFERENCES

1. Neshan, H. (1985). Water Hammer, pumps Iran Co. Tehran, Iran, pp. 1–60.

2. Streeter, V. L. and Wylie, E. B. (1979). *Fluid Mechanics*. McGraw-Hill Ltd., USA, pp. 492–505.

3. Hariri, K. (2007a). Decreasing of unaccounted for water "UFW" by Geographic Information System "GIS" in Rasht Urban Water System. Technical and Art. *J. Civil Engineering Organization of Gilan* **38**, 3–7.

4. Hariri, K. (2008b). GIS and water hammer disaster at earthquake in Rasht water pipeline. 3rd International Conference on Integrated Natural Disaster Management (INDM) .Tehran, Iran. Technical and Art. *J. Civil Engineering Organization of Gilan* 14–17.

5. Hariri, K. (2007c). Interpenetration of two fluids at parallel between plates and turbulent moving in pipe. 8th Conference on Ministry of Energetic works at research week. Tehran, Iran.

6. Hariri, K. (2007d). Water hammer and valves. 8th Conference on Ministry of Energetic works at research week. Tehran, Iran.

7. Hariri, K. (2007e). Water hammer and hydrodynamics instability. 8th Conference on Ministry of Energetic works at research week. Tehran, Iran.

8. Hariri, K. (2007f). Water hammer analysis and formulation. 8th Conference on Ministry of Energetic works at research week. Tehran, Iran.

9. Hariri, K. (2007g). Water hammer and fluid condition. 8th Conference on Ministry of Energetic works at research week. Tehran, Iran.

10. Hariri, K. (2007h). Water hammer and pump pulsation. 8th Conference on Ministry of Energetic works at research week. Tehran, Iran.

11. Hariri, K. (2007i). Reynolds number and hydrodynamics' instability. 8th Conference on Ministry of Energetic works at research week. Tehran, Iran.

12. Wylie, E. B. and Streeter, V. L. (1982). Fluid Transients, Feb Press, Ann Arbor, MI, (1983) corrected copy: 166–171.

13. Arturo, S. Leon (2007). An efficient second-order accurate shock-capturing scheme for modeling one and two-phase water hammer flows. Ph. D. *Thesis*, pp. 4–44.

4 Water Hammer and Hydrodynamics' Instability

CONTENTS

NOMENCLATURES

λ = coefficient of combination,

t = time,

$\rho 1$ = density of the light fluid (kg/m3),

$\rho 2$ = density of the heavy fluid (kg/m3),

s = length,

τ = shear stress,

C=surge wave velocity (m/s),

v2-v1=velocity difference (m/s),

e=pipe thickness (m),

K=module of elasticity of water(kg/m²),

C=wave velocity(m/s),

u = velocity (m/s),

D = diameter of each pipe (m),

θ = mixed ness integral measure,

R=pipe radius (m²),

w = weight

$\lambda_{,}$ = unit of length

V = velocity

C = surge wave velocity in pipe

f = friction factor

H2-H1=pressure difference (m-H2O)

g=acceleration of gravity (m/s²)

V=volume

Ee=module of elasticity(kg/m²)

θ = mixed ness integral measure

σ = viscous stress tensor

c = speed of pressure wave (celerity-m/s)

f = Darcy–Weisbach friction factor

μ=fluid dynamic viscosity(kg/m.s)

γ= specific weight (N/m³)

J=junction point (m),   I= moment of inertia (m^4)

A = pipe cross-sectional area (m²)   r= pipe radius (m)

d=pipe diameter(m), dp =is subjected to a static pressure rise

Ev=bulk modulus of elasticity,   α =kinetic energy correction factor

P=surge pressure (m),   ρ= density (kg/m3)

C = velocity of surge wave (m/s),   g=acceleration of gravity (m/s²)

ΔV= changes in velocity of water (m/s),   K = wave number

Tp = pipe thickness (m),   Ep = pipe module of elasticity (kg/m2)

Ew = module of elasticity of water (kg/m2),   C1=pipe support coefficient

T=time (s),   Ψ= depends on pipeline support-characteristics and Poisson's ratio

4.1   INTRODUCTION

The majority of transients in water and wastewater systems are the result of changes at system boundaries, typically at the upstream and downstream ends of the system or at local high points. Consequently, results of present chapter can reduce the risk of system damage or failure with proper analysis to determine the system's default dynamic response, design protection equipment to control transient energy, and specify operational procedures to avoid transients. Analysis, design, and operational procedures all benefit from computer simulations in this chapter. The study of hydraulic transients is generally considered to have begun with the works of Joukowski (1898) and Allievi (1902). The historical development of this subject makes for good reading. A number of pioneers made breakthrough contributions to the field, including R. Angus and John Parmakian (1963) and Wood (1970), who popularized and refined the graphical calculation method. Benjamin Wylie and Victor Streeter (1993) combined the method of characteristics with computer modeling. The field of fluid transients is still rapidly evolving worldwide by Brunone et al. (2000); Koelle and Luvizotto, (1996); Filion and Karney, (2002); Hamam and McCorquodale, (1982); Savic and Walters, (1995); Walski and Lutes, (1994); Wu and Simpson, (2000). Various methods have been developed to solve transient flow in pipes. These ranges have been formed from approximate equations to numerical solutions of the nonlinear Navier-Stokes equations. Water hammer uncontrolled energy appears as pressure spikes. Vibration and interpenetration between the water flows and mixture components is the visible example of water hammer and is the culprit that usually leads the way to component failure. A pump's motor exerts a torque on a shaft that delivers energy to the pump's impeller, forcing it to rotate and add energy to the fluid as it passes from the suction to the discharge side of the pump volute. Pumps convey fluid to the downstream end of a system whose profile can be either uphill or downhill, with irregularities such as local high or low points. When the pump starts, pressure can increase rapidly. Whenever power sags or fails, the pump slows or stops and a sudden drop in pressure propagates downstream (a rise in pressure also propagates upstream in the suction system). The similarity of the transient conditions caused by different source devices provides the key to transient analysis in a wide range of different systems. The initial state of the system and the ways in which energy and mass are added or removed from it must be considered. This is best illustrated by an example for a typical pumping system. In this chapter, a scale model which can be built to reproduce transients observed in a prototype (real) system have been compared by Field Tests. Water transmission line has been equipped with high-speed data loggers and Programmable Logic Control "PLC". Beside these metering instruments, advanced flow and pressure sensors have been installed in water pipeline (transmission line).

This method provided a suitable way for detecting, analysis and records of transient flow (down to 5 milliseconds). Then transient flow has been solved for pipeline in the range of approximate equations. These approximate equations have been solved by numerical solutions of the nonlinear Navier-Stokes equations using Method of Characteristics (MOC).

Long-distance water transmission lines must be economical, reliable and expandable. So, research was retained to provide hydraulic input to a network-wide optimization and risk-reduction strategy for Rasht city main pipeline. The track record includes multi-booster pressurized lines with surge protection ranging from check valves to gas vessels (surge tank). The water hammer phenomena investigation due to fluid condition was the main approach of this research. It formed by investigation of relation between: P-surge pressure (as a function or dependent variable with nomenclature "Y") and several factors (as independent variables with nomenclature "X") such as: ρ–density(kg/m³), C–velocity of surge wave (m/s), g–acceleration of gravity (m/s²), ΔV–changes in velocity of water (m/s), d–pipe diameter (m), T–pipe thickness (m), Ep–pipe module of elasticity (kg/m²), Ew–module of elasticity of water (kg/m²), C1–pipe support coefficient, T–Time (sec), Tp–pipe thickness (m). Water Hammer Software Version 07.00.049.00 results have been compared with Regression software "SPSS" as well. In the final procedure a Condition Base Maintenance (CM) method has been founded for all water transmission systems [1].

4.2 MATERIALS AND METHODS

The MOC have been widely used and tested. It is applied for complex boundary conditions and friction and vaporous cavitations models [2–5].

Equations (2.1) and (2.2) can not be solved analytically (Figure 4.1–4.2) .

$$\frac{1}{A}\left(\frac{dA}{dt}\right) - c^2\left(\frac{\partial v}{\partial s}\right) + \left(\frac{1}{\rho}\right)\left(\frac{1p}{dt}\right) = 0 \qquad \text{(Continuity equation), (4.1)}$$

$$\left(\frac{dv}{dt}\right) + \left(\frac{1}{\rho}\right)\left(\frac{\partial p}{\partial s}\right) + g\left(\frac{dz}{ds}\right) + \left(\frac{f}{2D}\right)V\left|V\right| = 0 \qquad \text{(Euler equation), (4.2)}$$

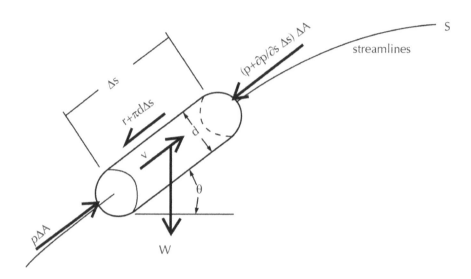

FIGURE 4.1   Newton second law-forces analysis for fluid element.

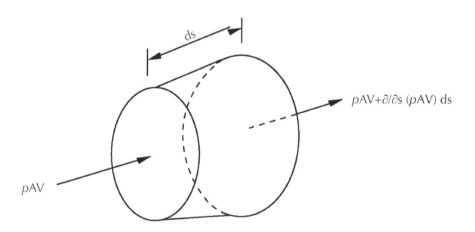

FIGURE 4.2   Continuity equation or conservation of momentum.

Those Equations can be expressed graphically in space-time as characteristic lines. Theses lines are called characteristics lines. (Figure 4.3–4.5) Theses lines represent signals propagating to the right (C+) and to the left (C–) simultaneously and from each location in the system.

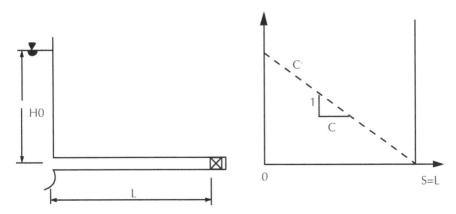

FIGURE 4.3   Coordination of (s-t).

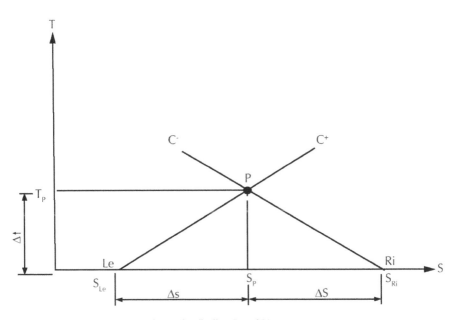

FIGURE 4.4   Coordination of (s-t) for finding P and V.

At each interior solution point, signals arrive from the two adjacent points simultaneously [2-5]. A linear combination of H and V are invariant along each characteristic if friction losses are neglected. Therefore, H and V can be obtained exactly at solution points. Head losses concentrated at solution points. The assumption is the friction is small. Transient modeling essentially consists of solving these equations. These equations are applied for every solution point and time step. It is valid for a wide variety of boundary conditions and system topologies.

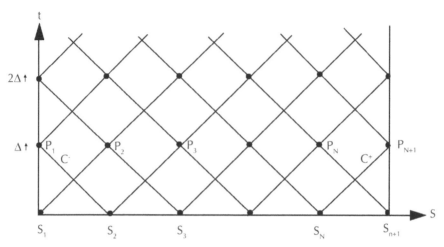

FIGURE 4.5   Characteristic lines network for assumed pipe.

4.2.1   Modeling and numerical simulations of water hammer

Two fundamental laws apply to steady state, or transient models:

- Conservation of mass—also expressed as the continuity equation, which states that matter cannot be created or destroyed.
- Conservation of energy—also expressed as the momentum equation, which states that energy cannot be created or destroyed. The best way to arrive at sound, physically meaningful conclusions and recommendations is to keep these principles in mind whenever research interprets the results of a hydraulic model. Present research has made this easy by tracking the mass inflow or outflow of air or water at any location and by plotting or animating the resulting total energy at any point and time in the system. Friction factor at steady state was the same as unsteady state [6-10]:

$$\tau_0 = \frac{\rho f v^2}{8}, \; -\left(\frac{1}{\gamma}\right)\left(\frac{\partial z}{\partial s}\right) - \left(\frac{f}{D}\right)\left(\frac{V^2}{2g}\right) = \left(\frac{1}{g}\right)\left(\frac{dV}{dt}\right), \tag{4.3}$$

For flow direction changes:

$$V^2 = V|V|, \tag{4.4}$$

$$\left(\frac{dV}{dt}\right) + \left(\frac{1}{\rho}\right)\left(\frac{\partial p}{\partial s}\right) + g\left(\frac{dz}{ds}\right) + \left(\frac{f}{2D}\right)V|V| = 0 \qquad \text{(Euler equation),} \tag{4.5}$$

Continuity equation and for fluid fine element:

$$\left(\frac{1}{\rho}\right)\left(\frac{d\rho d\rho}{d+1/A}\right)\left(\frac{dA}{dt}\right) - \left(\frac{\partial V}{\partial s}\right) + \left(\frac{1}{ds}\right)\left(\frac{d}{dt}\right)(ds) = 0, \tag{4.6}$$

-Fluid elasticity property,

$$\left(\frac{1}{\rho}\right)\left(\frac{d\rho\rho}{d}\right) = 0$$

-Pipe elasticity property

$$\frac{1}{A}\left(\frac{dA}{dt}\right) - c^2\left(\frac{\partial v}{\partial s}\right) + \left(\frac{1}{\rho}\right)\left(\frac{dp}{dt}\right) = 0 \qquad \text{(Continuity equation)}, \qquad (4.7)$$

$$\left(\frac{dV}{dt}\right) + \left(\frac{1}{\rho}\right)\left(\frac{\partial p}{\partial s}\right) + g\left(\frac{dz}{ds}\right) + \left(\frac{f}{2D}\right)V|V| = 0 \quad \text{(Euler equation)}, \qquad (4.8)$$

The Method of Characteristic (MOC), to solve governing equations for unsteady pipe flow. Using the MOC, the two partial differential equations can be transformed to the following two pairs of equations:

$$\frac{g}{c}\left(\frac{dH}{dt}\right) + \left(\frac{dV}{dt}\right) + \left(\frac{fV|V|}{2d}\right) = 0 \Rightarrow \frac{ds}{dt} = c^+ \qquad (4.9)$$

$$-\frac{g}{c}\left(\frac{dH}{dt}\right) + \left(\frac{dV}{dt}\right) + \left(\frac{fV|V|}{2D}\right) = 0 \Rightarrow \frac{ds}{dt} = c^- \qquad (4.10)$$

Method of characteristic solution for partial differential equation:

The method of characteristics is a finite difference technique where pressures are computed along the pipe for each time step. Calculation automatically sub-divides the pipe into sections (i.e. reaches or intervals) and selects a time interval for computations.

$$\left(\frac{dp}{dt}\right) = \left(\frac{\partial p}{\partial t}\right) = \left(\frac{\partial p}{\partial s}\right)\left(\frac{ds}{dt}\right) \qquad (4.11)$$

$$\left(\frac{dv}{dt}\right) = \left(\frac{\partial v}{\partial t}\right) + \left(\frac{\partial v}{\partial s}\right)\left(\frac{ds}{dt}\right) \qquad (4.12)$$

P and V changes due to time are high and due to coordination are low then we can neglect coordination differentiation, we have:

$$\left(\frac{\partial v}{\partial t}\right) + \left(\frac{1}{\rho}\right)\left(\frac{\partial p}{\partial s}\right) + g\left(\frac{dz}{ds}\right) + \left(\frac{f}{2D}\right)V|V| = 0 \qquad \text{(Euler equation)}, \quad (4.13)$$

$$C^2\left(\frac{\partial v}{\partial t}\right) + \left(\frac{1}{\rho}\right)\left(\frac{\partial p}{\partial t}\right) = 0 \qquad \text{(Continuity equation)}, \quad (4.14)$$

By linear combination of Euler and continuity equations in characteristic solution Method we have:,

$$\lambda\left[\frac{\partial v}{\partial s}+\left(\frac{1}{\rho}\right)\left(\frac{\partial p}{\partial s}\right)+g\left(\frac{dz}{ds}\right)+\left(\frac{f}{2D}\right)V|V|\right]+C^2\left(\frac{\partial V}{\partial s}\right)+\left(\frac{1}{\rho}\right)\left(\frac{\partial p}{\partial t}\right)=0, \lambda =^+ c \,\&\, \lambda =^- c \tag{4.15}$$

$$\left(\frac{dV}{dt}\right)+\left(\frac{1}{c\rho}\right)\left(\frac{dp}{dt}\right)+g\left(\frac{dz}{ds}\right)+\left(\frac{f}{2D}\right)V|V|=0 \tag{4.16}$$

$$\left(\frac{dV}{dt}\right)-\left(\frac{1}{c\rho}\right)\left(\frac{dp}{dt}\right)+g\left(\frac{dz}{ds}\right)\left(\frac{f}{2D}\right)V|V|=0 \tag{4.17}$$

Method of characteristics drawing in (s-t) coordination:

$$\left(\frac{dV}{dt}\right)-\left(\frac{g}{c}\right)\left(\frac{dH}{dt}\right)=0 \tag{4.18}$$

$$dH=\left(\frac{c}{g}\right)gx \qquad \text{(Joukowski Formula)}, \tag{4.19}$$

By Finite Difference method:

$$c+:\left(\frac{(V_P-V_{Le})}{(T_P-0)}\right)+\left(\left(\frac{g}{c}\right)\frac{(H_P-H_{Le})}{(t_P-0)}\right)+\left(\frac{fV_{Le}|V_{Le}|}{2D}\right)=0|, \tag{4.20}$$

$$c-:\left(\frac{V_P-V_{Ri}}{t_P-0}\right)+\left(\frac{g}{c}\right)\left(\frac{H_P-H_{Ri}}{t_P-0}\right)+\left(\frac{fV_{Ri}|V_{Ri}|}{2D}\right)=0|, \tag{4.21}$$

$$c+:(V_P-V_{Le})+\left(\frac{g}{c}\right)(H_P-H_{Le})+f\Delta f\left(\frac{fV_{Le}|V_{Le}|}{2D}\right)=0|, \tag{4.22}$$

$$c-:(V_P-V_{Ri})+\left(\frac{g}{c}\right)(H_P-H_{Ri})+f\Delta f\left(\frac{V_{Ri}|V_{Ri}|}{2D}\right)=0|, \tag{4.23}$$

The MOC approach transforms the water hammer partial differential equations into the ordinary differential equations along the characteristic lines. Theses lines defined as the continuity equation and the momentum equation are needed to determine V and P in a one-dimensional flow system. Solving these two equations produces a theoretical result that usually corresponds quite closely to actual system measurements. This is happened when the data and assumptions used to build the numerical model are valid. Transient analysis results that are not comparable with actual system measurements are generally caused by inappropriate system data (especially boundary conditions) and inappropriate assumptions. The MOC is based on a finite difference technique where pressures are computed along the pipe for each time step. [11]

$$H_{P=} \frac{1}{2}\left(\frac{C}{g}(V_{Le}-V_{ri})+(H_{Le}+H_{ri})-\frac{C}{g}\left(\frac{f\Delta t}{2D}\right)\left(V_{Le}\left|-V_{ri}\right|V_{ri}\right|\right)\right) \qquad (4.24)$$

$$V_{P}=\frac{1}{2}\left((V_{Le}+V_{ri})+\left(\frac{g}{c}\right)(H_{Le}-H_{ri})-\left(\frac{f\Delta t}{2D}\right)\left(V_{Le}\left|V_{Le}\right|+\left|V_{ri}\right|V_{ri}\right|\right)\right) \qquad (4.25)$$

f=friction, C=slope (deg.), V=velocity, t=time, H=head (m)

This research has provided Regression model for relation between P (water hammer surge pressure) and the related parameters for the Rasht city main pipeline, in the north of Iran.

Laboratory Models—a scale model have been built to reproduce transients observed in a prototype (real) system, typically for forensic or steam system investigations. This research Lab. model has recorded flow and pressure data (Table 4.1). The model is calibrated using one set of data and, without changing parameter values, it is used to match a different set of results. If successful, it is considered valid for these cases. (Figure 4.6)

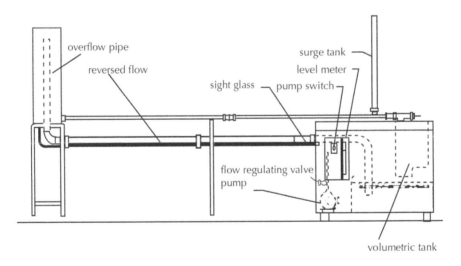

FIGURE 4.6   Research Laboratory Model for Laboratory experiments with flow and pressure data records.

The model has been calibrated and final checked by water hammer Laboratory Models of Iran science and Technology University in Tehran at: 05/02/09 with the flowing specifications: (Table 4.1).

TABLE 4.1 Research Laboratory Model Technical specifications.

Laboratory Model Technical specifications	notation	value	dimension
pipe diameter	d	22	mm
surge tank cross section area	A	$1.521*10^{-3}$	m²
pipe cross section area	a	$.3204*10^{-3}$	m²
pipe thickness	t	0.9	mm
fluid density	ρ	1000	kg/ m³
volumetric coefficient	K	2.05	GN/ m²
fluid power	P	*	*
fluid force	F	*	*
friction loss	hf	*	*
frequency	W	*	*
fluid velocity	v	*	m/s
Max fluctuation	Ymax	*	*
flow rate	q	*	m³/s
pipe length	L	*	m
period of motion	T	*	*
Surge tank and reservoir elevation difference	y	*	m
surge wave velocity	C	*	m/s

* Laboratory experiments and Field Tests results [12]

Field Tests—Field tests have provided key modeling parameters such as the pressure-wave speed or pump inertia. Water pipeline has equipped with advanced flow and pressure sensors, high-speed data loggers and "PLC". Hence fast transients, down to 5 milliseconds have been recorded. Methods such as inverse transient calibration and leak detection in calculation of Unaccounted for Water "UFW" have used such data. Field tests have been formed on actual systems with flow and pressure data records. These comparisons require threshold and span calibration of all sensor groups, multiple simultaneous datum and time base checks and careful test planning and interpretation [6-8].

Regression Equations—There is a relation between two or many Physical Units of variables. For example, there is a relation between volume of gases and their internal temperatures. The main approach in this research is investigation of relation between P = surge pressure(m)—as a function "Y"—and several factors—as variables "X"—such as; ρ–density (kg.m⁻³), C–velocity of surge wave (m.s⁻¹), g–acceleration of gravity (m.s⁻²), ΔV–changes in velocity of water (m.s⁻¹), d–pipe diameter (m), T–pipe thickness (m), Ep–pipe module of elasticity (kg.m⁻²), Ew–module of elasticity of water (kg.m⁻²), C1–pipe support coefficient, T–Time (sec), Tp–pipe thickness (m) [13].

4.2.2 Structure

There are case two cases for Modeling:

1. The combined elasticity of both the water and the pipe walls is characterized by the pressure wave speed (Arithmetic method combination of Joukowski Formula and Allievi Formula):

$$H_2 - H_1 = \left(\frac{C}{g}\right)(V_2 - V_1) = \rho C(V_2 - V_1),$$ (4.26) (Joukowski Formula)

$$C = \frac{1}{\left[\rho\left(\frac{1}{k}\right) + \left(\frac{dC_1}{Ee}\right)\right]1/2},$$ (4.27) (Allievi Formula)

With combination of Joukowski Formula and Allievi Formula:

$$H_2 - H_1 = 1/[\rho/((1/k) + (d.C1/E.e))]1/2(V_2 - V_1)/g,$$ (4.28)

Hence, water hammer pressure or surge pressure (ΔH) is a function of independent variables (**X**) such as:

$$\Delta H \approx \rho, K, d, C1, Ee, V, g,$$ (4.29)

2. The MOC based on a finite difference technique where pressures are computed along the pipe for each time step,

$$V_p = \frac{1}{2}\left((V_{Le} + V_{ri}) + \left(\frac{g}{c}\right)(H_{Le} - H_{ri}) - \left(\frac{f\Delta t}{2D}\right)(V_{Le}|V_{Le}| + V_{ri}|V_{ri}|)\right)$$ (4.30)

$$H_p = \frac{1}{2}\left(\frac{C}{g}(V_{Le} - V_{ri}) + (H_{Le} + H_{ri}) - \frac{C}{g}\left(\frac{f\Delta t}{2D}\right)(V_{Le}|V_{Le}| - V_{ri}|V_{ri}|)\right)$$ (4.31)

Hence, water hammer pressure or surge pressure (ΔH) is a function of independent variables (X) such as:

$$\Delta H \approx f, T, C, V, g, D,$$ (4.32)

For a model definition in this research relation between surge pressure (m)—as a function "Y"—and several factors—as variables "X"—have been investigated then Water hammer software evaluates transient flow data as a function of parameters: ρ, K, d, C1, Ee, V, f, T, C, g,Tp. Regression software have fitted the function curve and provides regression analysis.

Note: For simplicity and in the curve fitting process of function by regression software, assumed that H is dependent variable and V is the only independent variable. Hence, other variables assumed constant. [14]

4.3   RESULTS AND DISSCUSIONS

Like all fluid research, however, data are obtained at a finite number of locations and generalizing the findings requires assumptions, with uncertainties spread across the system.

TABLE 4.2   Model Summary and Parameter Estimates (Water hammer condition).

	Model	Un-standardized Coefficients	Standardized Coefficients		t	Sig.
			Std. Error	Beta		
1	(Constant)	28.762	29.73	-	0.967	0.346
	flow	0.031	0.01	0.399	2.944	0.009
	distance	-0.005	0.001	-0.588	-4.356	0
	time	0.731	0.464	0.117	1.574	0.133
2	(Constant)	14.265	29.344	-	0.486	0.632
	flow	0.036	0.01	0.469	3.533	0.002
	distance	-0.004	0.001	-0.52	-3.918	0.001
3	(Constant)	97.523	1.519	-	64.189	0
4	(Constant)	117.759	2.114	-	55.697	0
	distance	-0.008	0.001	-0.913	-10.033	0
5	(Constant)	14.265	29.344	-	0.486	0.632
	flow	0.036	0.01	0.469	3.533	0.002
	distance	-0.004	0.001	-0.52	-3.918	0.001

Regression Equation defined in stage (1) has been accepted, because its coefficients are meaningful:

$$\text{pressure} = 28.762 + .031\ \text{flow} - .005\ \text{Dostane} + .731\ \text{Time} \qquad (4.33)$$

At worst cases, tests can lead to physically doubtful conclusions limited by the scope of the test program. Neither laboratory models nor field testing can substitute for the careful and correct application of a proven hydraulic transient computer model. If a system is faced to large changes in velocity and pressure in short time periods, then transient analysis is required.

4.3.1   Regression modeling results has been compared with research Field Tests Model

Assumption: $p = f(V, T, L)$, V–velocity (flow), T–time, and L–distance are the most important requested variables. Regression software fitted the function curve (Figure 4.7) and have provided regression analysis. Results are shown in Table 4.1 "Model Summary and Parameter Estimates" (Table 4.2).

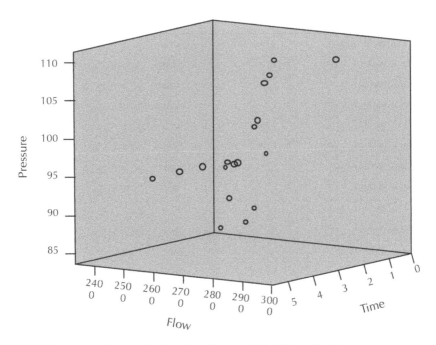

FIGURE 4.7 Scatter diagram for Lab Tests (Research Field Tests Model).

Research Field Tests Model
Software Hammer—Version 07.00.049.00
Type of Run: Full
Date of Run: 09/19/08
Time of Run: 04:47 am
Data File: E:\k-hariri Asli\ daraye nashti.inp
Hydrograph File: Not Selected
Labels: Short

TABLE 4.3 (a) System Information Preference, (b) Debug Information.

(a)	
ITEM	**VALUE**
Time Step (s)	
Automatic	0.0148
User-Selected	n/a
Total Number	339
Total Simulation Time (s)	5.00
Units	cms, m
Total Number of Nodes	27

TABLE 4.3   *(Continued)*

Total Number of Pipes	26
Specific Gravity	1.00
Wave Speed (m/s)	1084
Vapor Pressure (m)	-10.0
Reports	
Number of Nodes	52
Number of Time Steps	All
Number of Paths	1
Output	
Standard	No
Cavities (Open/Close)	No
Adjustments	
Adjusted Variable	Length
Warning Limit (%)	75.00
Calculate Transient Forces	No
Use Auxiliary Data File	Yes

(b)

ITEM	VALUE
Tolerances	
Initial Flow Consistency Value	0.0006
Initial Head Consistency Value	0.030
Criterion for Fr. Coef. Flag	0.025
Adjustments	
Elevation Decrease	0.00
Extreme Heads Display	
All Times	Yes
After First Extreme	No
Friction Coefficient	
Model	Steady
1,000,000 x Kinematic Viscosity	n/a
Debug Parameters	
Level	Null

TABLE 4.4   (a) The total paths, (b) The paths in 26 points, table Created by Hammer - Version 07.00.049.00 compared to equation of Regression software "SPSS".

a)

Path	No. of Points	From Point	To Point	Length (m)
1	249	P2:J6	P25:N1	3595.0

b)

Pipe	No. of Points	From Point	To Point	Length (m)
P2	5	P2:J6	P2:J7	60.7
P3	21	P3:J7	P3:J8	311.0
P4	2	P4:J3	P4:J4	1.0
P5	2	P5:J4	P5:J26	.5
P6	8	P6:J26	P6:J27	108.7
P7	3	P7:J27	P7:J6	21.5
P8	2	P8:J8	P8:J9	15.0
P9	23	P9:J9	P9:J10	340.7
P10	14	P10:J10	P10:J11	207.0
P11	22	P11:J11	P11:J12	339.0
P12	22	P12:J12	P12:J13	328.6
P13	4	P13:J13	P13:J14	47.0
P14	38	P14:J14	P14:J15	590.0
P15	4	P15:J15	P15:J16	49.0
P16	15	P16:J16	P16:J17	224.0
P17	2	P17:J17	P17:J18	18.4
P18	2	P18:J18	P18:J19	14.6
P19	2	P19:J19	P19:J20	12.0
P20	32	P20:J20	P20:J21	499.0
P21	17	P21:J21	P21:J22	243.4
P22	11	P22:J22	P22:J23	156.0
P23	3	P23:J23	P23:J24	22.0
P24	6	P24:J24	P24:J28	82.0
P25	4	P25:J28	P25:N1	35.6
P0	2	P0:J1	P0:J2	.5
P1	2	P1:J2	P1:J3	.5

TABLE 4.5 Pipe Information Created by Hammer—Version 07.00.049.00 (Snapshot for selected end points at time 4.9885 s) compared to equation of Regression software.

LABEL	LENGTH (m)	DIAMETER (mm)	WAVESPEED (m /s)	D-W FR. COEF.		CHECK VALVE
P2	60.7	1200.00	1084.0	0.028	*	
P3	311.0	1200.00	1084.0	0.028	*	
P4	1.0	1200.00	1084.0	0.027	*	
P5	0.5	1200.00	1084.0	0.027		
P6	108.7	1200.00	1084.0	0.067	*	
P7	21.5	1200.00	1084.0	0.028	*	
P8	15.0	1200.00	1084.0	0.028	*	
P9	340.7	1200.00	1084.0	0.028	*	
P10	207.0	1200.00	1084.0	0.028	*	
P11	339.0	1200.00	1084.0	0.028	*	
P12	328.6	1200.00	1084.0	0.028	*	
P13	47.0	1200.00	1084.0	0.028	*	
P14	590.0	1200.00	1084.0	0.028	*	
P15	49.0	1200.00	1084.0	0.028	*	
P16	224.0	1200.00	1084.0	0.028	*	
P17	18.4	1200.00	1084.0	0.028	*	
P18	14.6	1200.00	1084.0	0.028	*	
P19	12.0	1200.00	1084.0	0.028	*	
P20	499.0	1200.00	1084.0	0.028	*	
P21	243.4	1200.00	1084.0	0.028	*	
P22	156.0	1200.00	1084.0	0.028	*	
P23	22.0	1200.00	1084.0	0.036	*	
P24	82.0	1200.00	1084.0	0.028	*	
P25	35.6	1200.00	1084.0	0.028	*	
P0	0.5	1200.00	1084.0	0.027	*	
P1	0.5	1200.00	1084.0	0.027	*	

* Darcy-Weisbach friction coefficient exceeds .025

TABLE 4.6   Initial Pipe Conditions Information Created by Hammer—Version 07.00.049.00 (Snapshot for selected end points at time 4.9885 s) compared to equation of Regression software "SPSS".

FROM NODE	HEAD (m)	TO NODE	HEAD (m)	FLOW (cms)	VEL (m/s)
J6	133.4	J7	133.0	2.500	2.21
J7	133.0	J8	131.2	2.500	2.21
J3	135.0	J4	135.0	3.000	2.65
J4	135.0	J26	135.0	3.000	2.65
J26	135.0	J27	133.5	2.500	2.21
J27	133.5	J6	133.4	2.500	2.21
J8	131.2	J9	131.1	2.500	2.21
J9	131.1	J10	129.2	2.500	2.21
J10	129.2	J11	128.0	2.500	2.21
J11	128.0	J12	126.0	2.500	2.21
J12	126.0	J13	124.1	2.500	2.21
J13	124.1	J14	123.8	2.500	2.21
J14	123.9	J15	120.4	2.500	2.21
J15	120.4	J16	120.2	2.500	2.21
J16	120.2	J17	118.9	2.500	2.21
J17	118.9	J18	118.8	2.500	2.21
J18	118.8	J19	118.7	2.500	2.21
J19	118.7	J20	118.6	2.500	2.21
J20	118.6	J21	115.7	2.500	2.21
J21	115.7	J22	114.3	2.500	2.21
J22	114.3	J23	113.4	2.500	2.21
J23	113.5	J24	113.3	2.500	2.21
J24	113.3	J28	112.8	2.500	2.21
J28	112.8	N1	112.6	2.500	2.21
J1	40.6	J2	40.6	3.000	2.65
J2	40.6	J3	40.6	3.000	2.65

TABLE 4.7   Output data table Created by Hammer—Version 07.00.049.00 Min. & Max. head compared to equation of Regression software "SPSS".

END POINT	MAX. PRESS. (mH)	MIN. PRESS. (mH)	MAX. HEAD (m)	MIN. HEAD (m)
P2:J6	124.1	97.0	160.4	133.4
P2:J7	123.7	96.7	160.0	133.0
P3:J7	123.7	96.7	160.0	133.0
P3:J8	121.9	95.1	158.0	131.2

TABLE 4.7   *(Continued)*

END POINT	MAX. PRESS. (mH)	MIN. PRESS. (mH)	MAX. HEAD (m)	MIN. HEAD (m)
P4:J3	126.7	99.5	162.2	135.0
P4:J4	125.1	97.8	162.3	135.0
P5:J4	125.1	97.8	162.3	135.0
P5:J26	125.1	97.8	162.3	135.0
P6:J26	125.1	97.8	162.3	135.0
P6:J27	124.1	97.0	160.5	133.5
P7:J27	124.1	97.0	160.5	133.5
P7:J6	124.1	97.0	160.4	133.4
P8:J8	121.9	95.1	158.0	131.2
P8:J9	119.9	93.1	157.9	131.1
P9:J9	119.9	93.1	157.9	131.1
P9:J10	117.4	90.8	155.7	129.2
P10:J10	117.4	90.8	155.7	129.2
P10:J11	115.8	89.4	154.4	128.0
P11:J11	115.8	89.4	154.4	128.0
P11:J12	112.6	86.4	152.2	126.0
P12:J12	112.6	86.4	152.2	126.0
P12:J13	109.1	83.1	150.1	124.1
P13:J13	109.1	83.1	150.1	124.1
P13:J14	107.5	81.5	149.8	123.8
P14:J14	107.5	81.5	149.8	123.9
P14:J15	100.9	60.9	146.1	106.1
P15:J15	100.9	60.9	146.1	106.1
P15:J16	102.3	62.3	145.8	105.8
P16:J16	102.3	62.3	145.8	105.8
P16:J17	99.3	59.2	144.3	104.3
P17:J17	99.3	59.2	144.3	104.3
P17:J18	101.2	61.1	144.2	104.1
P18:J18	101.2	61.1	144.2	104.1
P18:J19	101.8	61.7	144.1	104.0
P19:J19	101.8	61.7	144.1	104.0
P19:J20	99.8	59.8	144.0	104.0
P20:J20	99.8	59.8	144.0	104.0
P20:J21	98.3	58.1	140.9	100.6
P21:J21	98.3	58.1	140.9	100.6
P21:J22	94.7	54.4	139.3	99.0

TABLE 4.7   *(Continued)*

| END | MAX. PRESS. | MIN. PRESS. | MAX. HEAD | MIN. HEAD |
POINT	(mH)	(mH)	(m)	(m)
P22:J22	94.7	54.4	139.3	99.0
P22:J23	68.4	28.1	138.3	98.0
P23:J23	68.4	28.1	138.3	98.0
P23:J24	56.3	15.9	138.1	97.7
P24:J24	56.3	15.9	138.1	97.7
P24:J28	42.4	0.0	137.6	95.2
P25:J28	42.4	0.0	137.6	95.2
P25:N1	16.7	16.7	112.6	112.6
P0:J1	0.0	0.0	40.6	40.6
P0:J2	27.5	2.1	63.0	37.6
P1:J2	27.5	2.1	63.0	37.6
P1:J3	27.5	2.1	63.0	37.6

Elapsed time: 12 s.

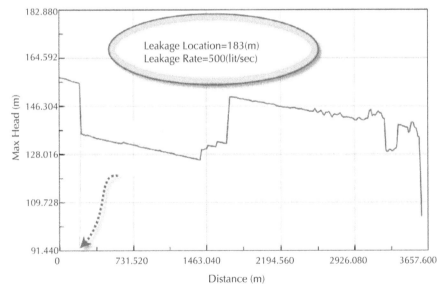

FIGURE 4.8   Rasht city Water Pipeline pilot Research (with surge tank and in leakage condition) Water flow decreased from 3000 (l/s) to 2500 (l/s).

4.4   CONCLUSION

In this chapter, Laboratory Model results are similar to a scale model which can be built to reproduce transients observed in a prototype (real) system. It has been compared by Field Tests results. Results indicates that at water hammer condition of (110M transient pressure), in the vicinity of pump station, leakage has been happened. Hence, the leakages of Transmission Line can occur nearby water treatment plant. Also water transmission flow decreased from 3000 (l/s) to 2500 (l/s).

Therefore, it is always a good idea to check extreme transient pressures for any system with large changes in elevation, long pipelines with large diameters (i.e., mass of water), and initial (e.g., steady-state) velocities in excess of 1 (m/s). In some cases, hydraulic transient forces can result in cracks or breaks, even with low steady-state velocities (Table 4.3–4.7).

KEYWORDS

- **Navier–Stokes equations**
- **Prototype system**
- **Pump volute**
- **Regression software**
- **Transient flow**
- **Water hammer software**

REFERENCES

1. Hariri, K. (2007). Interpenetration of two fluids at parallel between plates and turbulent moving in pipe. Ph. D. Thesis, pp. 8–12.
2. Streeter, V. L. and Lai, C. (1962). Water hammer Analysis Including Fluid Friction, *Journal of Hydraulics Division ASCE*, **88**, 79.
3. Chaudhry, M. H. (1987). Applied Hydraulic Transients. Van Nostrand Reinhold, New York.
4. Neshan, H. (1985). Water Hammer, pumps Iran Co. Tehran, Iran, pp. 1–60.
5. Streeter, V. L. and Wylie, E. B. (1979). Fluid Mechanics. McGraw-Hill Ltd., USA, pp. 492–505.
6. Hariri, K. (2007a). Decreasing of unaccounted for water "UFW" by Geographic Information System "GIS" in Rasht Urban Water System. Technical and Art. *J. Civil Engineering Organization of Gilan* **38**, 3–7.
7. Hariri, K. (2008b). GIS and water hammer disaster at earthquake in Rasht water pipeline. 3rd International Conference on Integrated Natural Disaster Management (INDM) .Tehran, Iran. Technical and Art *J. Civil Engineering Organization of Gilan*, 14–17.
8. Hariri, K. (2007c). Interpenetration of two fluids at parallel between plates and turbulent moving in pipe. 8th Conference on Ministry of Energetic works at research week. Tehran, Iran.
9. Hariri, K. (2007d). Water hammer and valves. 8th Conference on Ministry of Energetic works at research week. Tehran, Iran.
10. Hariri, K. (2007e). Water hammer and hydrodynamics' instability. 8th Conference on Ministry of Energetic works at research week. Tehran, Iran.
11. Hariri, K. (2007f). Water hammer analysis and formulation. 8th Conference on Ministry of Energetic works at research week. Tehran, Iran.
12. Hariri, K. (2007g). Water hammer and fluid condition. 8th Conference on Ministry of Energetic works at research week. Tehran, Iran.

13. Hariri, K. (2007h). Water hammer and pump pulsation. 8th Conference on Ministry of Energetic works at research week. Tehran, Iran.

14. Hariri, K. (2007). Interpenetration of two fluids at parallel between plates and turbulent moving in pipe. Ph. D. Thesis, pp. 8–12.

15. Wylie, E. B. and Streeter V. L. (1982). Fluid Transients, pp. 166–171.

5 Hadraulic Flow Control in Binary Mixtures

CONTENTS

NOMENCLATURES

R_0 = radiuses of a bubble
ρ_l = (consisting of components 1 and 2) binary mix with the density
k = mass concentration of a component of 1 mix
D = diffusion factor
β = cardinal influence of componential structure of a mixture is shown by value of the parameter
γ = adiabatic curve indicator
c_l, c_{pv} = accordingly specific thermal capacities of a liquid and pair at constant pressure a_l = factor thermal conductivity
ρ_v = steam density
t = time
R = vial radius
λ_l = heat conductivity factor
ΔT = a liquid overheat temperature

5.1 INTRODUCTION

Fluid interpenetration generally occurs when positive displacement takes place in a liquid motive force. It is most commonly caused by the acceleration and deceleration of the fluid. This uncontrolled energy appears as pressure spikes. Vibration and interpenetration between the water flows and mixture components is the visible example of pulsation and is the culprit that usually leads the way to component failure. A pump's motor exerts a torque on a shaft that delivers energy to the pump's impeller, forcing it to rotate and add energy to the fluid as it passes from the suction to the discharge side of the pump volute. Pumps convey fluid to the downstream end of a system whose profile can be either uphill or downhill, with irregularities such as local high or low points. When the pump starts, pressure can increase rapidly. Whenever power sags or fails, the pump slows or stops and a sudden drop in pressure propagates downstream. The similarity of the transient conditions caused by different source devices provides the key to transient analysis in a wide range of different systems: understand the initial state of the system and the ways in which energy and mass are added or removed from it. This is best illustrated by an example for typical binary mixtures [1-5].

As it is known, in various branches of manufacture there is a necessity of use of liquids for cooling of heating up surfaces. Thus, the reasons of allocation of a considerable quantity of heat can be various. For example, in engines of cars in the course of a friction and fuel ignition, at machine tool work, influence of a rotating drill on metal details and so on.

In all resulted examples, cooling surfaces is reached by means of influence of liquids. And for more effective cooling, binary mixtures (a mix of two liquids) are used. In many cases mixtures should be also cold-resistant, that is not to freeze. In language of mechanics it means that the liquid should have the minimum speed of phase transformations (for example, boiling up and freezing).

The binary mix with the density, ρ_l consisting of components 1 and 2, the resulted density accordingly was considered ρ_1 and ρ_2. And, $\rho_1 + \rho_2 = \rho_l$ $\rho_1 / \rho_l = k$, $\rho_2 / \rho_l = 1 - k$, where k = mass concentration of a component of 1 mix. The problem about radial movements of steam bubbles of a binary mixture at various pressure differences in a liquid and various initial radiuses of a bubble R_0 for water mixture of ethyl spirit representing to the big practical interest and ethylene glycol is solved for R_0 [4]. An interesting effect has been thus revealed. The parameters characterizing dynamics of bubbles in water mixture of ethyl spirit in the field of variable pressure lie between limiting values corresponding parameters for pure a component when pressure differences and accordingly a diffusion role are insignificant. At pressure difference increase along with thermal dissipation joins to diffusion dissipation. Thus, speed collapse and bubble growth considerably. Other situation is observed at growth and collapse a steam bubble in water mixture ethylene glycol. In this case the effect diffusion the resistance leading to braking of speed of phase transformations is observed.

5.2 MATERIALS AND METHODS

There are many possible causes for rapid or sudden changes in a pipe system, including power failures, pipe breaks, or a rapid valve opening or closure. These can result from natural causes, equipment malfunction, or even operator error. It is, therefore, important to consider the several ways in which hydraulic transients can occur in a system and to model them. If identifying, modeling, and protecting against several possible hydraulic transient events seem to take a lot of time and resources, remember that it is far safer and less expensive to learn about your system's vulnerabilities by "breaking pipes" in a computer model—and far easier to clean up—than from expensive service interruptions and field repairs. At pressure difference increase along with the interpenetration fluid condition is a combination of the diffusing process and remixing process. Thus, speed collapse and bubble growth considerably above. Influences of heat exchange and diffusion on decrement of attenuation of free fluctuations of a steam bubble of a binary mixture were investigated in literature [4-7]. It was found out that dependence of decrement of attenuation of fluctuations of a bubble of water mixture of ethyl spirit, methanol, toluene are monotonous on while k_0 for a water mixture ethylene glycol similar dependence has a characteristic minimum at $k_0 \approx 0,02$ [2]. And at decrement $0,01 \le k_0 \le 0,3$ of attenuation of a binary mixture there are less than decrements of attenuation of pulsations of a bubble in pure single component to water and ethylene glycol. It means that in ranges of concentration of water of a pulsation $0,01 \le k_0 \le 0,3$ of a bubble of a water mixture ethylene glycol fade much more slowly and there is a braking of process of phase transformations. The similar picture is revealed and at the compelled fluctuations of bubbles in an acoustic field [7]. The influence of non-stationary mass transfer processes on distribution of waves to a binary mixture of liquids with bubbles is reported in [8]. Influence of componential structure and concentration of a binary mixture on a dispersion, dissipation and attenuation of monochromatic waves in diphase, two-componential environments is analyzed. Unlike system of a water mixture of ethyl spirit in a water mixture ethylene glycol decrements of attenuation of indignations less than corresponding characteristics in pure components of a mixture. In work [9] calculate structure of stationary shock waves in bubbly binary mixtures taking into account non-stationary inter phase heat exchange. As show calculations, and in the problems considered in this work the effect of infringement of monotony of behavior of settlement curves on the concentration, testifying to presence diffusion resistance is shown. Apparently in all listed problems in a number of binary mixes the effect diffusion the resistance leading to braking of intensity of phase transformations is shown. The physical explanation of why in a water mixture ethylene glycol so the effect diffusion resistance is brightly shown consists that in such mixture at the limited ability a component diffusion through each other ($D = 10^{-9} m^2 / \sec$, D—diffusion factor) evaporation rate a component strongly differ, and as consequence. Very strongly concentration a component in a mixture and a steam phase differ. In a case of a water solution of ethyl spirit evaporation rate a component are approximately identical $\chi_1^0 \approx \chi_2^0$. Therefore, $c_0 \approx k_0$ and finiteness of factor of diffusion does not lead to essential effects at infringement of thermal and mechanical balance of phases. It is necessary to notice that in all works listed above irrespective of the considered problems at the mathematical description cardinal influence of componential structure of a mixture is shown by value of the parameter, β equal

$$\beta = \left(1 - \frac{1}{\gamma}\right) \frac{(c_0 - k_0)(N_{c_0} - N_{k_0})}{k_0(1 - k_0)} \frac{c_l}{c_p} \left(\frac{c_p T_0}{L}\right)^2 \sqrt{\frac{a_l}{D}}, \qquad (5.1)$$

Where N_{k_0}, N_{c_0} mole concentration of 1-th component in a liquid and steam

$$N_{k_0} = \frac{\mu k_0}{\mu k_0 + 1 - k_0}, \quad N_{c_0} = \frac{\mu c_0}{\mu c_0 + 1 - c_0}$$

γ = Adiabatic curve indicator, and c_l, c_{pv} = accordingly specific thermal capacities of a liquid and pair at constant pressure, a_l = factor thermal conductivity $L = l_1 c_0 + l_2(1 - c_0)$.

Let us notice also that the parameter (2.1) is included into the author modeling decision, a vial describing growth in superheated mixture. This decision looks like:

$$R = 2\sqrt{\frac{3}{\pi}} \frac{\lambda_l \Delta T \sqrt{t}}{L \rho_v \sqrt{a_l}(1 + \beta)} \tag{5.2}$$

Here, ρ_v = steam density, t = time, R = vial radius, λ_l = heat conductivity factor, ΔT = a liquid overheat. On Figure 5.1 and Figure 5.2 dependences for $\beta(k_0)$ considered above binary mixtures are resulted $\beta(k_0)$. For water mixture of ethyl spirit β is negative for any value of concentration and depends on k_0 monotonous character. For a water mixture ethylene glycol β—it is positive and has a strongly pronounced maximum at $k_0 = 0,02$. At small pressure differences (accordingly overheats and liquid overcooling) diffusion is not shown in water mixtures of ethyl spirit. In view of approximate equality and k_0 all c_0 settlement dependences lies between limiting curves for a case of single components of a mixture.

5.3 RESULTS AND DISCUSSION

By pressure difference increasing, essential diffusion processes between a bubble and a liquid has been increased. By pressure difference increasing, mass transfer between a bubble and a liquid has been increased. Comparison with pure water and ethyl spirit, growths rate of a bubble have been happened according to β (2.2) decreasing. In ethylene glycol water mixture small indignations between k_0 and c_0 (especially at $0,01 \leq k_0 \leq 0,3$) has been accounted. So, the effect of diffusion braking has been promoted essential delay of mass transfer intensity. Compare with pure water and ethylene glycol, at increasing of β (2.2) in a mixture, growth of a bubble was very low. β Maximum value (at $k_0 = 0,02$) have been led to maximum effect of braking. The similar condition has been observed at pulsations and collapse of a bubble. Similar to work [6] attenuation decrement dependence of pulsations had monotonous character. This has not been observed in a water mixture of ethyl spirit from concentration of water. It had a minimum at $k_0 = 0,02$. So, at value $0,01 \leq k_0 \leq 0,3$ decrement of attenuation was small. The difference was accordingly great between k_0 and c_0 so it accepted β great values. Therefore, in these ranges of concentration of a mixture the effect diffusion braking was essential.

5.4 CONCLUSION

For glycerin, methanol, toluene experimental data has been showed the effect of essential delay or braking of processes in warmly mass transfer. In binary mixtures dependence of parameter β bubble pulsations attenuation from equilibrium concentration of a mix component have been analyzed. This has been defined concentration a component of a binary mixture. Figure 5.3 and Figure 5.4 have presented dependences $k_0(c_0)$ for water mixtures of ethyl spirit and ethylene glycol. Figure 5.4 has been showed interval change ($k_0, k_0 \approx c_0$). So for ethylene glycol water mixture, from calculations (Figure 5.5) has been showed at $0,01 \le k_0 \le 0,3$ for $k_0 \le c_0$, and at $k_0 > 0,3$ for $k_0 \sim c_0$.

For considered above systems from Figure 5.5 and Figure 5.6 boiling temperature dependences on concentration of a mixture have been resulted. At $k_0 = 1$ and $c_0 = 1$ pure water with steam bubble has been received. Temperature of boiling of such liquid was ($T_0 = 373^0 K$). At $k_0 = 0$ and $c_0 = 0$ accordingly pure bubbly ethyl spirit $(T_0 = 350^0 K)$ ethylene glycol $(T_0 = 470^0 K)$ has been received.

KEYWORDS

- **Diffusion dissipation**
- **Ethyl spirit**
- **Ethylene glycol**
- **Monochromatic waves**
- **Thermal dissipation**

REFERENCES

1. Hariri Asli, K. (2007). Interpenetration of two fluids at parallel between plates and turbulent moving in pipe. Ph. D. Thesis, pp. 8-12.

2. Vargaftic, N. B. (1972). The directory of thermo physics properties of gases and liquids. M, "Nauka".

3. Gerasimov, Y. I. (1963). A physical chemistry course. Vol.1. M Goskhimizdat.

4. Nagiyev, F. B. and Kadirov, B. A., (1985). Heat exchange and dynamics of steam bubbles in two-component mixture of liquids. Proceedings Academy of Sciences Azerbaijan, N 4, pp. 10-13.

5. Khabeev, N. S. and Nagiyev, F. B., (1989). Dynamics of bubbles in mixtures. Thermodynamics of high temperature (TVT) 27, N3, pp. 528-533.

6. Nagiyev, F. B. (1988). Attenuation of oscillations of bubble boiling binary mixtures. Proceedings of VIII conference for mathematics and mechanics, Baku, pp. 177–178.

7. Nagiyev, F. B. and Kadirov, B. A. (1986). Small oscillations of bubbles of two-component mixture in acoustic field. Bulletin Academy of Sciences Azerbaijan, ser. phys.-tech. math. sc., N1, pp. 150-153.

8. Nagiyev, F. B. (1986). Linear theory of propagation of waves in binary bubbly mixture of liquids. Dep. In VINITI 17.01.86., N 405, V 86.

9. Nagiyev, F. B. (1989). The structure of stationary shock waves in boiling binary mixtures. Bulletin Academy of Sciences USSR, Mechanics of Liquid and Gas (MJG), N1, pp. 81-87.

6 An Efficient Accurate Shock-capturing Scheme for Modeling Water Hammer Flows

CONTENTS

NOMENCLATURES

λ = coefficient of combination,

t = time,

$\rho 1$ = density of the light fluid (kg/m3),

$\rho 2$ = density of the heavy fluid (kg/m3),

s = length,

τ = shear stress,

C = surge wave velocity (m/s),

$v2$-$v1$ = velocity difference (m/s),

e = pipe thickness (m),

K = module of elasticity of water(kg/m²),

w = weight

$\lambda_.$ = unit of length

V = velocity

C = surge wave velocity in pipe

f = friction factor

$H2$-$H1$ = pressure difference (m-H2O)

g = acceleration of gravity (m/s²)

V = volume

Ee = module of elasticity(kg/m²)

θ = mixed ness integral measure

C = wave velocity(m/s),

u = velocity (m/s),

D = diameter of each pipe (m),

θ = mixed ness integral measure,

R = pipe radius (m²),

J = junction point (m),

A = Pipe cross-sectional area (m²)

d = pipe diameter(m),

Ev = bulk modulus of elasticity,

P = surge pressure (m),

C = velocity of surge wave (m/s),

ΔV = changes in velocity of water (m/s),

Tp = pipe thickness (m),

Ew = module of elasticity of water (kg/m2),

T = time (s),

σ = viscous stress tensor

c = speed of pressure wave (celerity-m/s)

f = Darcy–Weisbach friction factor

μ = fluid dynamic viscosity(kg/m.s)

γ = specific weight (N/m³)

I = moment of inertia (m^4)

r= pipe radius (m)

dp = is subjected to a static pressure rise

α = kinetic energy correction factor

ρ = density (kg/m3)

g = acceleration of gravity (m/s²)

K = wave number

Ep = pipe module of elasticity (kg/m2)

C1 = pipe support coefficient

Ψ = depends on pipeline support-characteristics and Poisson's ratio

6.1   INTRODUCTION AND OVERVIEW

On-Line analysis of transient flow and evaluating Programmable Logic Control (PLC) input-output by Geography Information System (GIS) in order to automatic control of water pipeline and surge tank, for condition base maintenance "CM" is the scientific novelty of present research. The pressure transducer which is very sensitive, has a high resolution, and is connected to a high-speed data acquisition unit. It is also connected to the system pipeline with a device to release air, because air can distort the pressure signal transmitted during the transient. Recording will not begin until all air is released from the pipeline connection and the pressure measurement interval is defined. Typically, at least two measuring locations are established in the system and the flow-control operation is closely monitored. The timings of all recording equipment are synchronized. For valves, the movement of the position indicator is recorded as a function of time. For pumps, rotation or speed is measured over time. For protection devices such as surge tanks (hydro-pneumatic tanks), the level is measured over time. Advanced flow and pressure sensors equipped with high-speed data loggers and "PLC" in water pipeline makes it possible to capture fast transients, down to 5 ms for interpenetration between water flows. Changes in fluid properties such as depressurization due to the sudden opening of a relief valve, a propagating pressure pulse, heating or cooling in cogeneration or industrial systems, mixing with solids or other liquids may affect fluid density, specific gravity, and viscosity can generate fast transients. So formation and collapse of vapor bubbles (cavitations), and air entrainment or release from the system (at air vents and/or due to pressure waves) can generate fast transients. Changes at system boundaries such as rapidly opening or closing a valve, pipe burst (due to high pressure) or pipe collapse (due to low pressure), pump start/shift/stop, air intake at a vacuum breaker can generate fast transients. Water intake at a valve, mass outflow at a pressure-relief valve or fire hose, breakage of a rupture disk, and hunting and/or resonance at a control valve can generate fast transients. Sudden changes such as these create a transient pressure pulse that rapidly propagates away from the disturbance, in every possible direction, and throughout the entire pressurized system (Figure 6.1).

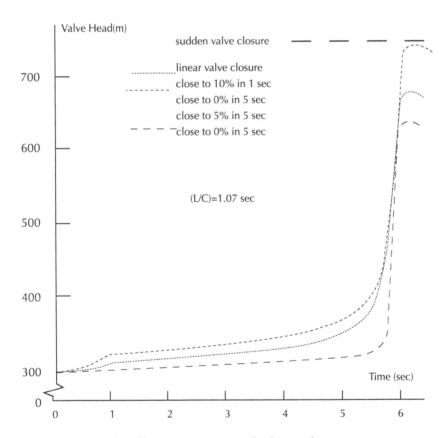

FIGURE 6.1   Valve closing effect on pressure generation in water hammer.

If no other transient event is triggered by the pressure wave fronts, unsteady-flow conditions continue until the transient energy is completely damped and dissipated by friction [1-2]. This chapter have two objectives: (1) to make the hydraulic engineers aware of the system conditions that lead to the development of undesirable transients, such as pump and valve operations, and (2) to present the protection methods and devices that should be used during design and construction of particular systems and discuss their practical limitations. Pipeline geo reference coordination can be collected in database under GIS. By data exchange management between receiver and transducer on pipeline and PLC, all of the system can be on-line controlled. By transmitting pulse to valves, pump station and sure tank they can be protected from water hammer disaster. Unaccounted for Water (UFW) is main effect of water hammer disaster. So decreasing of UFW is the main practical aim of present chapter.

6.2   MATERIALS AND METHODS

Sudden deceleration of a check valve may slam shut rapidly, depending on the dynamic characteristic of the check valve and the mass of the water between a check valve and tank. This may occur after full pumping station trip or during pump trip of a single pump (last one is often more critical as downstream pressure is maintained, but first one may be as critical if a large air vessel is present). Resonance of appendages (mainly pumps) is a frequent cause of vibrations in the pipeline. Other fittings (bends etc) may sometimes induce vibrations due to shedding of vortices). These fluctuations are often of limited amplitude and will only cause problems if their frequency is close to natural frequencies in the pipeline. A valve can start, change, or stop flow very suddenly. Energy conversions increase or decrease in proportion to a valve's closing or opening rate and position, or stroke. Orifices can be used to throttle flow instead of a partially open valve. Valves can also allow air into a pipeline and/or expel it, typically at local high points [3-4]. Suddenly closing a flow-control valve (with piping on both sides) generates transients on both sides of the valve, as follows:

- Water initially coming towards the valve suddenly has nowhere to go. As water packs into a finite space upstream of the valve, it generates a high-pressure pulse that propagates upstream, away from the valve (Figure 6.2).
- Water initially going away from the valve cannot suddenly stop, due to its inertia and, since no flow is coming through the valve to replace it, the area downstream of the valve may "pull a vacuum", causing a low-pressure pulse to propagate downstream(Figures 6.3).

6.2.1   Possible Causes

Sudden release of air from the system (fire hydrant, air release valve) causes the fluid to suddenly stop moving at the location of the device (as the flow rate is suddenly set to zero (in case of non controlled or oversized air valve) or because there is a much larger head drop for the fluid than for the air (fire hydrant).

- *Field Tests*—Field tests can provide key modeling parameters such as the pressure-wave speed or pump inertia. Advanced flow and pressure sensors equipped with high-speed data loggers and "PLC" in water pipeline makes it possible to capture fast transients, down to 5 ms. Methods such as inverse transient calibration and leak detection in calculation of UFW use such data.

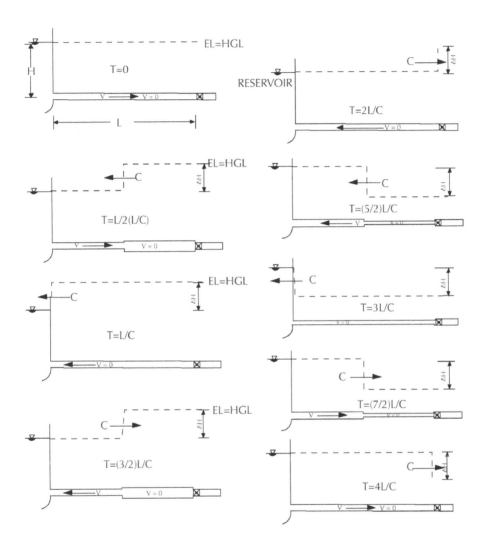

FIGURE 6.2   Surge wave motion.

Results of present chapter can reduce the risk of system damage or failure with proper analysis to determine the system's default dynamic response, design protection equipment to control transient energy, and specify operational procedures to avoid transients.

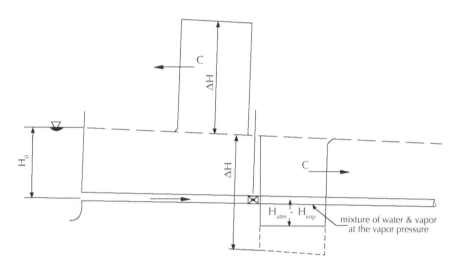

FIGURE 6.3   Column separations due to close of valve.

In present research data collection process, the main pipeline of water networks of Rasht city in the north of Iran have been used (Research Field Tests Model). Model of main pipeline of networks have been selected for the city of Rasht, Guilan province (with 1,050,000 population). Research data were collected from the PLC of water treatment plant pump stations. Water hammer models of the Rasht city pipeline guided route selection, conceptual and detailed design of pipeline. The pipeline is included water treatment plant pump station (in the start of water transmission), 3.595 km of 2*1.200 mm diameter pre-stressed pipes and one 50,000 m³ reservoir (at the end of water transmission). All of these parts have been tied into existing water networks.

6.2.2   Modeling and Simulations of Water Hammer

Two fundamental laws apply to steady state, or transient models:

- Conservation of mass—also expressed as the continuity equation, which states that matter cannot be created or destroyed.
- Conservation of energy—also expressed as the momentum equation, which states that energy cannot be created or destroyed. The best way to arrive at sound, physically meaningful conclusions and recommendations is to keep these principles in mind whenever research has interpreted the results of a hydraulic model. Present research has made this easy by tracking the mass inflow or outflow of air or water at any location. This goal has been achieved by plotting or animating the resulted total energy at any point and time in the system. Friction factor at steady state was the same as unsteady state [4-6]:

$$\tau_0 = \frac{\rho f v^2}{8}, -\left(\frac{1}{\gamma}\right)\left(\frac{\partial z}{\partial s}\right) - \left(\frac{\partial z}{\partial s}\right) - \left(\frac{f}{D}\right)\left(\frac{V^2}{2g}\right) = \left(\frac{1}{g}\right)\left(\frac{dV}{dt}\right), \tag{6.1}$$

For flow direction changes:

$$V^2 = V|V| \tag{6.2}$$

$$\left(\frac{dV}{dt}\right) + \left(\frac{1}{\rho}\right)\left(\frac{\partial p}{\partial s}\right) + g\left(\frac{dz}{ds}\right) + \left(\frac{f}{2D}\right)V|V| = 0, \text{ (Euler equation)} \tag{6.3}$$

Continuity equation and for fluid fine element:

$$\left(\frac{1}{\rho}\right)\frac{d\rho d\rho}{d} + \frac{1}{A}\left(\frac{dA}{Dt}\right) - \left(\frac{\partial V}{\partial s}\right) + \left(\frac{1}{ds}\right)\left(\frac{d}{dt}\right)(ds) = 0 \tag{6.4}$$

-fluid elasticity property,

$$\left(\frac{1}{\rho}\right)\left(\frac{d\rho\rho}{d}\right) = 0$$

-Pipe elasticity property

$$\frac{1}{A}\left(\frac{dA}{dt}\right) - c^2\left(\frac{\partial v}{\partial s}\right) + \left(\frac{1}{\rho}\right)\left(\frac{dp}{dt}\right) = 0, \text{ (Continuity equation)}, \tag{6.5}$$

$$\left(\frac{dV}{dt}\right) + \left(\frac{1}{\rho}\right)\left(\frac{\partial p}{\partial s}\right) + g\left(\frac{dz}{ds}\right) + \left(\frac{f}{2D}\right)V|V| = 0, \text{ (Euler Equation)}, \tag{6.6}$$

Using the MOC, the two partial differential equations can be transformed to the following equations:

$$\frac{g}{c}\left(\frac{dH}{dt}\right) + \frac{dV}{dt} + \left(fv|v|/2d\right) = 0 \Rightarrow \frac{ds}{dt} = c^+, \tag{6.7}$$

$$-\frac{g}{c}\left(\frac{dH}{dt}\right) + \frac{dV}{dt} + \left(fV|V|/2D\right) = 0 \Rightarrow \frac{ds}{dt} = c^-, \tag{6.8}$$

Method of characteristic solution for partial differential equation:
The method of characteristics (MOC) is a finite difference technique where pressures are computed along the pipe for each time step:

$$\left(\frac{dp}{dt}\right) = \left(\frac{\partial p}{\partial t}\right) + \left(\frac{\partial p}{\partial s}\right)\left(\frac{ds}{dt}\right) \tag{6.9}$$

$$\left(\frac{dv}{dt}\right) = \left(\frac{\partial v}{\partial t}\right) + \left(\frac{\partial v}{\partial s}\right)\left(\frac{ds}{dt}\right) \tag{6.10}$$

P and V changes due to time are high and due to coordination are low then we can neglect coordination differentiation:

$$\left(\frac{\partial v}{dt}\right) + \left(\frac{1}{\rho}\right)\left(\frac{\partial p}{\partial s}\right) + g\left(\frac{dz}{ds}\right) + \left(\frac{f}{2D}\right)V|V| = 0 \text{ (Euler equation)}, \tag{6.11}$$

$$C^2\left(\frac{\partial v}{dt}\right) + \left(\frac{1}{\rho}\right)\left(\frac{\partial p}{\partial t}\right) = 0 \text{ (Continuity equation)}, \tag{6.12}$$

By linear combination of Euler and continuity equations in characteristic solution method we have:

$$\lambda\left[\left(\frac{\partial V}{\partial t}\right) + \left(\frac{1}{\rho}\right)\left(\frac{\partial p}{\partial s}\right) + g\left(\frac{dz}{ds}\right) + \left(\frac{f}{2D}\right)V|V|\right] + C^2\left(\frac{\partial V}{\partial s}\right) + \left(\frac{1}{\rho}\right)\left(\frac{\partial p}{\partial t}\right) = 0 \tag{6.13}$$

$$\lambda = {}^+ c \,\&\, \lambda = {}^- c$$

$$\left(\frac{dV}{dt}\right) + \left(\frac{1}{c\rho}\right)\left(\frac{dp}{dt}\right) + g\left(\frac{dz}{ds}\right) + \left(\frac{f}{2D}\right)V|V| = 0, \tag{6.14}$$

$$\left(\frac{dV}{dt}\right) - \left(\frac{1}{c\rho}\right)\left(\frac{dp}{dt}\right) + g\left(\frac{dz}{ds}\right) + \left(\frac{f}{2D}\right)V|V| = 0, \tag{6.15}$$

The MOC drawing in (s-t) coordination:

$$\left(\frac{dV}{dt}\right) - \left(\frac{g}{c}\right)\left(\frac{dH}{dt}\right) = 0, \tag{6.16}$$

$$dH = \left(\frac{c}{g}\right)dv, \text{ (Joukowski Formula)}, \tag{6.17}$$

By Finite Difference method:

$$c+:\frac{\left(V_p - V_{Le}\right)}{\left(t_p - 0\right)} + \left(\left(\frac{g}{c}\right)\frac{\left(H_p - H_{Le}\right)}{\left(t_p - 0\right)}\right) + \left(\left(fV_{Le}|V_{Le}|\right)/2D\right) = 0,|, \tag{6.18}$$

$$c-:\frac{\left(V_p - V_{Ri}\right)}{\left(t_p - 0\right)} + \left(\left(\frac{g}{c}\right)\frac{\left(H_p - H_{Ri}\right)}{\left(t_p - 0\right)}\right) + \left(\left(fV_{Ri}|V_{Ri}|\right)/2D\right) = 0,|, \tag{6.19}$$

$$c+ : \left(V_p - V_{Le}\right) + \left(\frac{g}{c}\right)\left(H_p - H_{Le}\right) + \left(f\Delta f\left(fV_{Le}\left|V_{Le}\right|\right)/2D\right) = 0,| , \qquad (6.20)$$

$$c- : \left(V_p - V_{Ri}\right) + \left(\frac{g}{c}\right)\left(H_p - H_{Ri}\right) + \left(f\Delta f\left(V_{Ri}\left|V_{Ri}\right|\right)/2D\right) = 0,| , \qquad (6.21)$$

Solving these equations produced a theoretical result that usually corresponds quite closely to actual system measurements (if the data and assumptions used to build the numerical model are valid). Transient analysis results that are not comparable with actual system measurements are generally caused by inappropriate system data (especially boundary conditions) and inappropriate assumptions. The MOC is based on a finite difference technique where pressures are computed along the pipe for each time step [7-8]:

$$H_p = \frac{1}{2}\left(\frac{C}{g}\left(V_{Le} - V_{ri}\right) + \left(H_{Le} + H_{ri}\right) - \frac{c}{g}\left(f\frac{\Delta t}{2D}\right)\left(V_{Le}\left|V_{Le}\right| - V_{ri}\left|V_{ri}\right|\right)\right), \qquad (6.22)$$

$$V_p = \frac{1}{2}\left(\left(V_{Le} - V_{ri}\right) + \left(\frac{g}{c}\right)\left(H_{Le} - H_{ri}\right) - \left(f\frac{\Delta t}{2D}\right)\left(V_{Le}\left|V_{Le}\right| - V_{ri}\left|V_{ri}\right|\right)\right), \qquad (6.23)$$

f = friction, C = slope (deg.), V = velocity, t = time, H = head (m)

6.3 RESULTS AND DISCUSSION

Water hammer pressure or surge pressure (ΔH) is a function of independent variables (X) such as: $\Delta H \approx \rho$, C1, Ep, E w, V, T, C, g, Tp f, g, D, L. Hence, this chapter calibrated and validated numerical simulations for three parameter: p = f (V, T, L). Input data were in relation to water hammer condition.

TABLE 6.1   Model Summary and Parameter Estimates (Water hammer condition).

Model		Un-standardized Coefficients		Standardized Coefficients	t	Sig.
		B	Std. Error	Beta		
1	(Constant)	28.762	29.730		.967	.346
	flow	.031	.010	.399	2.944	.009
	distance	-.005	.001	-.588	-4.356	.000
	time	.731	.464	.117	1.574	.133
2	(Constant)	14.265	29.344		.486	.632
	flow	.036	.010	.469	3.533	.002
	distance	-.004	.001	-.520	-3.918	.001
3	(Constant)	97.523	1.519		64.189	.000
4	(Constant)	117.759	2.114		55.697	.000
	distance	-.008	.001	-.913	-10.033	.000

TABLE 6.1   *(Continued)*

Model		Un-standardized Coefficients		Standardized Coefficients	t	Sig.
		B	**Std. Error**	**Beta**		
5	(Constant)	14.265	29.344		.486	.632
	flow	.036	.010	.469	3.533	.002
	distance	-.004	.001	-.520	-3.918	.001

Regression Equation defined in stage (1) can be applied, because its coefficients are meaningful:

$$\text{pressure} = 28.762 + .031 \text{ Flow} - .005 \text{ Distane} + .731 \text{ Time} \tag{3.1}$$

6.3.1   Regression Analysis:

Assumption: $p = f (V, T, L)$, V—velocity (flow) and T—time and L—distance are the most important requested variables. Regression software has fitted the function curve (Figures 6.4) and provided regression analysis.

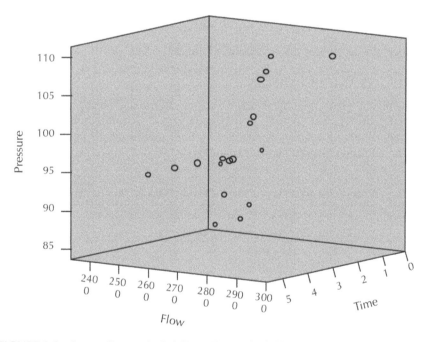

FIGURE 6.4   Scatter diagram for Lab Tests (Research Field Tests Model).

This chapter referred to a fluid transient as a "Dynamic" operating case, which may also include pulsation and sudden thrust due to relief valves that pop open or rapid piping accelerations (due to an earthquake). It is advisable to investigate fluid-structure interpenetration (FSI). This model shows the relation between two or many of variables accordance to fluid transient as a "Dynamic" operating (Table 6.1).

Rapidly closing or opening a valve causes pressure transients in pipelines, known as water hammer. Valve closure can result in pressures well over the steady state values, while valve opening can cause seriously low pressures, possibly so low that the flowing liquid vaporizes inside the pipe. Results showed the maximum and minimum piezometric pressures (relative to atmospheric) in each pipe in a pipeline as well as the time and location at which they have occurred [9]. The analysis is helpful in design to determine the maximum (or minimum) expected pressures due to valve closure or opening. By investigating the cause of pipe rupture, the model have provided to what rate of UFW (Figures 6.4–6.5). So it has defined location that water leakage has been happened on water pipeline due to the water hammer pressure [10-11].

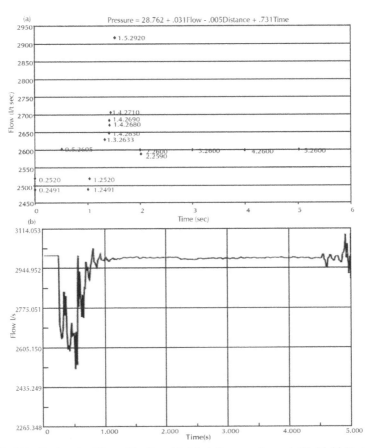

FIGURE 6.5 Rasht city Water Pipeline (a) Modeling results and (b) Field Tests results, comparison).

Research Field Tests Model (water pipeline of Rasht city in the north of Iran)
Software. Hammer—Version 07.00.049.00

Type of Run: Full
Date of Run: 09/19/08
Time of Run: 04:47 am
Data File: E:\k-hariri Asli\ daraye nashti.inp
Hydrograph File: Not Selected
Labels: Short

TABLE 6.2   (a) System information preference, (b) Debug information.

(a)	
ITEM	**VALUE**
Time Step (s)	
Automatic	0.0148
User-Selected	n/a
Total Number	339
Total Simulation Time (s)	5.00
Units	cms, m
Total Number of Nodes	27
Total Number of Pipes	26
Specific Gravity	1.00
Wave Speed (m/s)	1084
Vapor Pressure (m)	-10.0
Reports	
Number of Nodes	52
Number of Time Steps	All
Number of Paths	1
Output	
Standard	No
Cavities (Open/Close)	No
Adjustments	
Adjusted Variable	Length
Warning Limit (%)	75.00
Calculate Transient Forces	No
Use Auxiliary Data File	Yes
(b)	
ITEM	**VALUE**
Tolerances	
Initial Flow Consistency Value	0.0006
Initial Head Consistency Value	0.030
Criterion for Fr. Coef. Flag	0.025

TABLE 6.2   *(Continued)*

(b)	
ITEM	**VALUE**
Adjustments	
Elevation Decrease	0.00
Extreme Heads Display	
All Times	Yes
After First Extreme	No
Friction Coefficient	
Model	Steady
1,000,000 x Kinematic Viscosity	n/a
Debug Parameters	
Level	Null

TABLE 6.3   (a) The total paths, (b) The paths in 26 points, table Created by Hammer—Version 07.00.049.00 compared to equation of Regression software "SPSS".

(a)				
No. of Path	From Points	To Point	Length	(m)
1	249	P2:J6	P25:N1	3595.0

(b)				
Pipe	No. of Points	From Point	To Point	Length (m)
P2	5	P2:J6	P2:J7	60.7
P3	21	P3:J7	P3:J8	311.0
P4	2	P4:J3	P4:J4	1.0
P5	2	P5:J4	P5:J26	.5
P6	8	P6:J26	P6:J27	108.7
P7	3	P7:J27	P7:J6	21.5
P8	2	P8:J8	P8:J9	15.0
P9	23	P9:J9	P9:J10	340.7
P10	14	P10:J10	P10:J11	207.0
P11	22	P11:J11	P11:J12	339.0
P12	22	P12:J12	P12:J13	328.6
P13	4	P13:J13	P13:J14	47.0
P14	38	P14:J14	P14:J15	590.0
P15	4	P15:J15	P15:J16	49.0
P16	15	P16:J16	P16:J17	224.0
P17	2	P17:J17	P17:J18	18.4

TABLE 6.3   *(Continued)*

		(b)		
	No. of	From	To	Length
Pipe	Points	Point	Point	(m)
P18	2	P18:J18	P18:J19	14.6
P19	2	P19:J19	P19:J20	12.0
P20	32	P20:J20	P20:J21	499.0
P21	17	P21:J21	P21:J22	243.4
P22	11	P22:J22	P22:J23	156.0
P23	3	P23:J23	P23:J24	22.0
P24	6	P24:J24	P24:J28	82.0
P25	4	P25:J28	P25:N1	35.6
P0	2	P0:J1	P0:J2	.5
P1	2	P1:J2	P1:J3	.5

TABLE 6.4   Pipe information created by Hammer—Version 07.00.049.00 (Snapshot for selected end points at time 4.9885 s) compared to equation of Regression software.

LENGTH	DIAMETER	WAVESPEED	D-W	CHECK
LABEL	(m)	(mm)	(m /s)	FR. COEF.	VALVE
P2	60.7	1200.00	1084.0	0.028	*
P3	311.0	1200.00	1084.0	0.028	*
P4	1.0	1200.00	1084.0	0.027	*
P5	0.5	1200.00	1084.0	0.027	*
P6	108.7	1200.00	1084.0	0.067	*
P7	21.5	1200.00	1084.0	0.028	*
P8	15.0	1200.00	1084.0	0.028	*
P9	340.7	1200.00	1084.0	0.028	*
P10	207.0	1200.00	1084.0	0.028	*
P11	339.0	1200.00	1084.0	0.028	*
P12	328.6	1200.00	1084.0	0.028	*
P13	47.0	1200.00	1084.0	0.028	*
P14	590.0	1200.00	1084.0	0.028	*
P15	49.0	1200.00	1084.0	0.028	*
P16	224.0	1200.00	1084.0	0.028	*
P17	18.4	1200.00	1084.0	0.028	*
P18	14.6	1200.00	1084.0	0.028	*
P19	12.0	1200.00	1084.0	0.028	*
P20	499.0	1200.00	1084.0	0.028	*
P21	243.4	1200.00	1084.0	0.028	*

TABLE 6.4   *(Continued)*

LENGTH LABEL	DIAMETER (m)	(mm)	WAVESPEED (m /s)	D-W FR. COEF.	CHECK VALVE
P22	156.0	1200.00	1084.0	0.028	*
P23	22.0	1200.00	1084.0	0.036	*
P24	82.0	1200.00	1084.0	0.028	*
P25	35.6	1200.00	1084.0	0.028	*
P0	0.5	1200.00	1084.0	0.027	*
P1	0.5	1200.00	1084.0	0.027	*

* Darcy-Weisbach friction coefficient exceeds .025.

TABLE 6.5   Initial pipe conditions information created by Hammer—Version 07.00.049.00 (Snapshot for selected end points at time 4.9885 s) compared to equation of Regression software "SPSS".

FROM NODE	HEAD (m)	TO NODE	HEAD (m)	FLOW (cms)	VEL (m/s)
J6	133.4	J7	133.0	2.500	2.21
J7	133.0	J8	131.2	2.500	2.21
J3	135.0	J4	135.0	3.000	2.65
J4	135.0	J26	135.0	3.000	2.65
J26	135.0	J27	133.5	2.500	2.21
J27	133.5	J6	133.4	2.500	2.21
J8	131.2	J9	131.1	2.500	2.21
J9	131.1	J10	129.2	2.500	2.21
J10	129.2	J11	128.0	2.500	2.21
J11	128.0	J12	126.0	2.500	2.21
J12	126.0	J13	124.1	2.500	2.21
J13	124.1	J14	123.8	2.500	2.21
J14	123.9	J15	120.4	2.500	2.21
J15	120.4	J16	120.2	2.500	2.21
J16	120.2	J17	118.9	2.500	2.21
J17	118.9	J18	118.8	2.500	2.21
J18	118.8	J19	118.7	2.500	2.21
J19	118.7	J20	118.6	2.500	2.21
J20	118.6	J21	115.7	2.500	2.21
J21	115.7	J22	114.3	2.500	2.21
J22	114.3	J23	113.4	2.500	2.21
J23	113.5	J24	113.3	2.500	2.21
J24	113.3	J28	112.8	2.500	2.21

TABLE 6.5   *(Continued)*

FROM NODE	HEAD (m)	TO NODE	HEAD (m)	FLOW (cms)	VEL (m/s)
J28	112.8	N1	112.6	2.500	2.21
J1	40.6	J2	40.6	3.000	2.65
J2	40.6	J3	40.6	3.000	2.65

TABLE 6.6   Output data table created by Hammer—Version 07.00.049.00 Min. and Max. head compared to equation of Regression software "SPSS".

END POINT	MAX. PRESS (mH)	MIN. PRESS (mH)	MAX. HEAD (m)	MIN. HEAD (m)
P2:J6	124.1	97.0	160.4	133.4
P2:J7	123.7	96.7	160.0	133.0
P3:J7	123.7	96.7	160.0	133.0
P3:J8	121.9	95.1	158.0	131.2
P4:J3	126.7	99.5	162.2	135.0
P4:J4	125.1	97.8	162.3	135.0
P5:J4	125.1	97.8	162.3	135.0
P5:J26	125.1	97.8	162.3	135.0
P6:J26	125.1	97.8	162.3	135.0
P6:J27	124.1	97.0	160.5	133.5
P7:J27	124.1	97.0	160.5	133.5
P7:J6	124.1	97.0	160.4	133.4
P8:J8	121.9	95.1	158.0	131.2
P8:J9	119.9	93.1	157.9	131.1
P9:J9	119.9	93.1	157.9	131.1
P9:J10	117.4	90.8	155.7	129.2
P10:J10	117.4	90.8	155.7	129.2
P10:J11	115.8	89.4	154.4	128.0
P11:J11	115.8	89.4	154.4	128.0
P11:J12	112.6	86.4	152.2	126.0
P12:J12	112.6	86.4	152.2	126.0
P12:J13	109.1	83.1	150.1	124.1
P13:J13	109.1	83.1	150.1	124.1
P13:J14	107.5	81.5	149.8	123.8
P14:J14	107.5	81.5	149.8	123.9
P14:J15	100.9	60.9	146.1	106.1
P15:J15	100.9	60.9	146.1	106.1
P15:J16	102.3	62.3	145.8	105.8

TABLE 6.6   *(Continued)*

END POINT	MAX. PRESS (mH)	MIN. PRESS (mH)	MAX. HEAD (m)	MIN. HEAD (m)
P16:J16	102.3	62.3	145.8	105.8
P16:J17	99.3	59.2	144.3	104.3
P17:J17	99.3	59.2	144.3	104.3
P17:J18	101.2	61.1	144.2	104.1
P18:J18	101.2	61.1	144.2	104.1
P18:J19	101.8	61.7	144.1	104.0
P19:J19	101.8	61.7	144.1	104.0
P19:J20	99.8	59.8	144.0	104.0
P20:J20	99.8	59.8	144.0	104.0
P20:J21	98.3	58.1	140.9	100.6
P21:J21	98.3	58.1	140.9	100.6
P21:J22	94.7	54.4	139.3	99.0
P22:J22	94.7	54.4	139.3	99.0
P22:J23	68.4	28.1	138.3	98.0
P23:J23	68.4	28.1	138.3	98.0
P23:J24	56.3	15.9	138.1	97.7
P24:J24	56.3	15.9	138.1	97.7
P24:J28	42.4	0.0	137.6	95.2
P25:J28	42.4	0.0	137.6	95.2
P25:N1	16.7	16.7	112.6	112.6
P0:J1	0.0	0.0	40.6	40.6
P0:J2	27.5	2.1	63.0	37.6
P1:J2	27.5	2.1	63.0	37.6
P1:J3	27.5	2.1	63.0	37.6

Elapsed time: 12 s.

TABLE 6.7   Output data table created by Hammer—Version 07.00.049.00 (Snapshot for selected end points at time 0.0000 s) compared to equation of Regression software "SPSS".

End Point	Flow (cms)	Head (m)	Volume3 (m3)
P0:J1	3.000	40.6	.000
P0:J2	3.000	40.6	.000
P1:J2	3.000	40.6	.000
P1:J3	3.000	40.6	.000
P10:J10	3.000	127.4	.000
P10:J11	3.000	125.8	.000
P11:J11	3.000	125.7	.000

TABLE 6.7   *(Continued)*

End Point	Flow (cms)	Head (m)	Volume3 (m3)
P11:J12	3.000	123.0	.000
P12:J12	3.000	123.0	.000
P12:J13	3.000	120.4	.000
P13:J13	3.000	120.3	.000
P13:J14	3.000	120.0	.000
P14:J14	3.000	119.9	.000
P14:J15	3.000	115.2	.000
P15:J15	3.000	115.2	.000
P15:J16	3.000	114.9	.000
P16:J16	3.000	114.8	.000
P16:J17	3.000	113.0	.000
P17:J17	3.000	113.0	.000
P17:J18	3.000	112.9	.000
P18:J18	3.000	112.8	.000
P18:J19	3.000	112.8	.000
P19:J19	3.000	112.7	.000
P19:J20	3.000	112.7	.000
P2:J6	3.000	133.3	.000
P2:J7	3.000	132.8	.000
P20:J20	3.000	112.6	.000
P20:J21	3.000	108.6	.000
P21:J21	3.000	108.6	.000
P21:J22	3.000	106.7	.000
P22:J22	3.000	106.6	.000
P22:J23	3.000	105.4	.000
P23:J23	3.000	105.4	.000
P23:J24	3.000	105.2	.000
P24:J24	3.000	105.2	.000
P24:J28	3.000	104.6	.000
P25:J28	3.000	104.5	.000
P25:N1	3.000	104.3	.000
P3:J7	2.500	132.8	.000
P3:J8	2.500	130.3	.000
P4:J3	3.000	135.0	.000
P4:J4	3.000	135.0	.000
P5:J26	3.000	135.0	.000

TABLE 6.7   *(Continued)*

End	Flow	Head	Volume3
Point	(cms)	(m)	(m3)
P5:J4	3.000	135.0	.000
P6:J26	3.000	134.9	.000
P6:J27	3.000	133.5	.000
P7:J27	3.000	133.5	.000
P7:J6	3.000	133.3	.000
P8:J8	3.000	130.3	.000
P8:J9	3.000	130.2	.000
P9:J10	3.000	127.5	.000
P9:J9	3.000	130.1	.000

TABLE 6.8   Output data table created by Hammer—Version 07.00.049.00 (History for selected times at end point P9:J9) compared to equation of Regression software.

Time	Flow	Head	Volume 3
(s)	(cms)	(m)	(m3)
0.0000	3.000	130.1	.000
0.0148	3.000	130.1	.000
0.0295	2.750	106.5	.000
0.0443	2.750	106.5	.000
0.0590	2.750	106.4	.000
0.0738	2.750	106.4	.000
0.0886	2.750	106.4	.000
0.1033	2.750	106.4	.000
0.1181	2.749	106.4	.000
0.1328	2.749	106.4	.000
0.1476	2.749	106.4	.000
0.1623	2.749	106.4	.000
0.1771	2.749	106.4	.000
0.1919	2.749	106.4	.000
0.2066	2.749	106.3	.000
0.2214	2.749	106.3	.000
0.2361	2.749	106.3	.000
0.2509	2.749	106.3	.000
0.2657	2.749	106.3	.000
0.2804	2.749	106.3	.000
0.2952	2.749	106.3	.000
0.3099	2.749	106.3	.000

TABLE 6.8   *(Continued)*

Time	Flow	Head	Volume 3
(s)	(cms)	(m)	(m3)
0.3247	2.994	129.3	.000
0.3395	2.994	129.3	.000
0.3542	2.994	129.4	.000
0.3690	2.994	129.4	.000
0.3837	2.994	129.3	.000
0.3985	2.994	129.3	.000
0.4133	2.994	129.3	.000
0.4280	2.994	129.3	.000
0.4428	2.951	125.4	.000
0.4575	2.951	125.4	.000
0.4723	2.951	125.4	.000
0.4870	2.951	125.4	.000
0.5018	2.995	129.5	.000
0.5166	2.995	129.5	.000
0.5313	2.995	129.5	.000
0.5461	2.995	129.5	.000
0.5608	2.996	129.6	.000
0.5756	2.996	129.6	.000
0.5904	2.996	129.6	.000
0.6051	2.996	129.6	.000
0.6199	2.999	129.9	.000
0.6346	2.999	129.9	.000
0.6494	2.999	129.9	.000
0.6642	2.999	129.9	.000
0.6789	3.003	129.6	.000
0.6937	3.003	129.6	.000
0.7084	2.789	109.3	.000
0.7232	2.789	109.3	.000
0.7379	2.802	110.5	.000
0.7527	2.802	110.5	.000
0.7675	2.815	111.8	.000
0.7822	2.815	111.8	.000
0.7970	2.747	105.3	.000
0.8117	2.747	105.3	.000
0.8265	2.771	107.5	.000
0.8413	2.771	107.5	.000

TABLE 6.8   *(Continued)*

Time	Flow	Head	Volume 3
(s)	(cms)	(m)	(m3)
0.8560	2.747	105.2	.000
0.8708	2.747	105.2	.000
0.8855	2.767	107.2	.000
0.9003	2.767	107.2	.000
0.9151	2.711	101.8	.000
0.9298	2.711	101.8	.000
0.9446	2.720	102.7	.000
0.9593	2.720	102.7	.000
0.9741	2.738	104.9	.000
0.9889	2.738	104.9	.000
1.0036	2.963	126.2	.000
1.0184	2.963	126.2	.000
1.0331	2.981	127.9	.000
1.0479	2.981	127.9	.000
1.0626	2.986	128.1	.000
1.0774	2.986	128.1	.000
1.0922	2.998	129.1	.000
1.1069	2.998	129.1	.000
1.1217	2.983	127.7	.000
1.1364	2.983	127.7	.000
1.1512	2.989	128.3	.000
1.1660	2.989	128.3	.000
1.1807	2.995	128.9	.000
1.1955	2.995	128.9	.000
1.2102	3.040	133.2	.000
1.2250	3.040	133.2	.000
1.2398	3.036	132.8	.000
1.2545	3.036	132.8	.000
1.2693	3.005	129.9	.000
1.2840	3.005	129.9	.000
1.2988	3.002	129.6	.000
1.3136	3.002	129.6	.000
1.3283	3.002	129.6	.000
1.3431	3.002	129.6	.000
1.3578	2.994	128.6	.000
1.3726	2.994	128.6	.000

TABLE 6.8   *(Continued)*

Time	Flow	Head	Volume 3
(s)	(cms)	(m)	(m3)
1.3873	3.002	129.4	.000
1.4021	3.002	129.4	.000
1.4169	3.001	129.3	.000
1.4316	3.001	129.3	.000
1.4464	3.008	129.9	.000
1.4611	3.008	129.9	.000
1.4759	2.990	128.1	.000
1.4907	2.990	128.1	.000
1.5054	2.994	128.4	.000
1.5202	2.994	128.4	.000
1.5349	2.993	128.5	.000
1.5497	2.993	128.5	.000
1.5645	2.998	128.8	.000
1.5792	2.998	128.8	.000
1.5940	2.994	128.5	.000
1.6087	2.994	128.5	.000
1.6235	3.000	129.0	.000
1.6382	3.000	129.0	.000
1.6530	2.992	128.9	.000
1.6678	2.992	128.9	.000
1.6825	2.989	129.4	.000
1.6973	2.989	129.4	.000
1.7120	2.988	129.3	.000
1.7268	2.988	129.3	.000
1.7416	2.996	129.8	.000
1.7563	2.996	129.8	.000
1.7711	2.991	129.3	.000
1.7858	2.991	129.3	.000
1.8006	2.996	129.8	.000
1.8154	2.996	129.8	.000
1.8301	2.993	129.4	.000
1.8449	2.993	129.4	.000
1.8596	2.995	129.8	.000
1.8744	2.995	129.8	.000
1.8892	2.991	129.3	.000
1.9039	2.991	129.3	.000

TABLE 6.8   *(Continued)*

Time	Flow	Head	Volume 3
(s)	(cms)	(m)	(m3)
1.9187	2.994	129.6	.000
1.9334	2.994	129.6	.000
1.9482	2.994	129.5	.000
1.9629	2.994	129.5	.000
1.9777	2.999	129.3	.000
1.9925	2.999	129.3	.000
2.0072	2.994	128.9	.000
2.0220	2.994	128.9	.000
2.0367	2.994	129.1	.000
2.0515	2.994	129.1	.000
2.0663	2.994	129.2	.000
2.0810	2.994	129.2	.000
2.0958	2.995	129.4	.000
2.1105	2.995	129.4	.000
2.1253	2.994	129.3	.000
2.1401	2.994	129.3	.000
2.1548	2.996	129.3	.000
2.1696	2.996	129.3	.000
2.1843	2.995	129.3	.000
2.1991	2.995	129.3	.000
2.2139	2.995	129.2	.000
2.2286	2.995	129.2	.000
2.2434	2.993	129.1	.000
2.2581	2.993	129.1	.000
2.2729	2.995	129.3	.000
2.2876	2.995	129.3	.000
2.3024	2.994	129.2	.000
2.3172	2.994	129.2	.000
2.3319	2.995	129.2	.000
2.3467	2.995	129.2	.000
2.3614	2.992	129.6	.000
2.3762	2.992	129.6	.000
2.3910	2.995	129.4	.000
2.4057	2.995	129.4	.000
2.4205	2.993	129.2	.000
2.4352	2.993	129.2	.000

TABLE 6.8 *(Continued)*

Time	Flow	Head	Volume 3
(s)	(cms)	(m)	(m3)
2.4500	2.994	129.4	.000
2.4648	2.994	129.4	.000
2.4795	2.995	129.5	.000
2.4943	2.995	129.5	.000
2.5090	2.996	129.6	.000
2.5238	2.996	129.6	.000
2.5385	2.995	129.4	.000
2.5533	2.995	129.4	.000
2.5681	2.995	129.6	.000
2.5828	2.995	129.6	.000
2.5976	2.995	129.6	.000
2.6123	2.995	129.6	.000
2.6271	2.995	129.6	.000
2.6419	2.995	129.6	.000
2.6566	2.995	129.0	.000
2.6714	2.995	129.0	.000
2.6861	2.990	128.8	.000
2.7009	2.990	128.8	.000
2.7157	2.991	128.8	.000
2.7304	2.991	128.8	.000
2.7452	2.992	128.9	.000
2.7599	2.992	128.9	.000
2.7747	2.990	128.9	.000
2.7895	2.990	128.9	.000
2.8042	2.991	128.8	.000
2.8190	2.991	128.8	.000
2.8337	2.991	128.8	.000
2.8485	2.991	128.8	.000
2.8632	2.991	128.7	.000
2.8780	2.991	128.7	.000
2.8928	2.990	128.7	.000
2.9075	2.990	128.7	.000
2.9223	2.991	128.8	.000
2.9370	2.991	128.8	.000
2.9518	2.991	128.9	.000
2.9666	2.991	128.9	.000

TABLE 6.8   *(Continued)*

Time	Flow	Head	Volume 3
(s)	(cms)	(m)	(m3)
2.9813	2.995	129.2	.000
2.9961	2.995	129.2	.000
3.0108	2.994	129.2	.000
3.0256	2.994	129.2	.000
3.0404	2.994	129.1	.000
3.0551	2.994	129.1	.000
3.0699	2.995	128.9	.000
3.0846	2.995	128.9	.000
3.0994	2.994	128.8	.000
3.1141	2.994	128.8	.000
3.1289	2.994	128.8	.000
3.1437	2.994	128.8	.000
3.1584	2.996	128.9	.000
3.1732	2.996	128.9	.000
3.1879	2.996	128.9	.000
3.2027	2.996	128.9	.000
3.2175	2.996	128.9	.000
3.2322	2.996	128.9	.000
3.2470	2.995	128.8	.000
3.2617	2.995	128.8	.000
3.2765	2.995	128.8	.000
3.2913	2.995	128.8	.000
3.3060	2.995	128.8	.000
3.3208	2.995	128.8	.000
3.3355	2.995	128.9	.000
3.3503	2.995	128.9	.000
3.3651	2.991	128.8	.000
3.3798	2.991	128.8	.000
3.3946	2.993	129.0	.000
3.4093	2.993	129.0	.000
3.4241	2.993	129.0	.000
3.4388	2.993	129.0	.000
3.4536	2.991	128.9	.000
3.4684	2.991	128.9	.000
3.4831	2.995	128.7	.000
3.4979	2.995	128.7	.000

TABLE 6.8 *(Continued)*

Time	Flow	Head	Volume 3
(s)	(cms)	(m)	(m3)
3.5126	2.994	128.6	.000
3.5274	2.994	128.6	.000
3.5422	2.995	128.7	.000
3.5569	2.995	128.7	.000
3.5717	2.991	128.8	.000
3.5864	2.991	128.8	.000
3.6012	2.991	128.8	.000
3.6160	2.991	128.8	.000
3.6307	2.992	128.9	.000
3.6455	2.992	128.9	.000
3.6602	2.994	129.1	.000
3.6750	2.994	129.1	.000
3.6897	2.993	129.0	.000
3.7045	2.993	129.0	.000
3.7193	2.993	129.0	.000
3.7340	2.993	129.0	.000
3.7488	2.994	129.0	.000
3.7635	2.994	129.0	.000
3.7783	2.990	129.2	.000
3.7931	2.990	129.2	.000
3.8078	2.991	129.3	.000
3.8226	2.991	129.3	.000
3.8373	2.992	129.3	.000
3.8521	2.992	129.3	.000
3.8669	2.994	129.2	.000
3.8816	2.994	129.2	.000
3.8964	2.994	129.1	.000
3.9111	2.994	129.1	.000
3.9259	2.994	129.0	.000
3.9407	2.994	129.0	.000
3.9554	2.993	129.1	.000
3.9702	2.993	129.1	.000
3.9849	3.009	127.7	.000
3.9997	3.009	127.7	.000
4.0144	2.982	130.0	.000
4.0292	2.982	130.0	.000

TABLE 6.8   *(Continued)*

Time	Flow	Head	Volume 3
(s)	(cms)	(m)	(m3)
4.0440	2.962	132.0	.000
4.0587	2.962	132.0	.000
4.0735	2.995	129.3	.000
4.0882	2.995	129.3	.000
4.1030	2.995	129.2	.000
4.1178	2.995	129.2	.000
4.1325	2.995	129.1	.000
4.1473	2.995	129.1	.000
4.1620	2.998	129.0	.000
4.1768	2.998	129.0	.000
4.1916	2.998	129.0	.000
4.2063	2.998	129.0	.000
4.2211	2.999	129.0	.000
4.2358	2.999	129.0	.000
4.2506	2.997	129.1	.000
4.2654	2.997	129.1	.000
4.2801	2.982	130.6	.000
4.2949	2.982	130.6	.000
4.3096	3.007	128.2	.000
4.3244	3.007	128.2	.000
4.3391	3.027	126.3	.000
4.3539	3.027	126.3	.000
4.3687	2.995	128.9	.000
4.3834	2.995	128.9	.000
4.3982	2.998	128.7	.000
4.4129	2.998	128.7	.000
4.4277	2.992	129.2	.000
4.4425	2.992	129.2	.000
4.4572	2.983	130.0	.000
4.4720	2.983	130.0	.000
4.4867	2.995	129.4	.000
4.5015	2.995	129.4	.000
4.5163	2.999	129.0	.000
4.5310	2.999	129.0	.000
4.5458	2.995	129.3	.000
4.5605	2.995	129.3	.000

TABLE 6.8   *(Continued)*

Time	Flow	Head	Volume 3
(s)	(cms)	(m)	(m3)
4.5753	2.993	129.1	.000
4.5900	2.993	129.1	.000
4.6048	2.994	129.1	.000
4.6196	2.994	129.1	.000
4.6343	2.995	129.1	.000
4.6491	2.995	129.1	.000
4.6638	3.006	127.8	.000
4.6786	3.006	127.8	.000
4.6934	2.984	130.0	.000
4.7081	2.984	130.0	.000
4.7229	2.969	131.9	.000
4.7376	2.969	131.9	.000
4.7524	2.999	128.4	.000
4.7672	2.999	128.4	.000
4.7819	2.981	128.3	.000
4.7967	2.981	128.3	.000
4.8114	2.987	129.5	.000
4.8262	2.987	129.5	.000
4.8410	2.998	129.3	.000
4.8557	2.998	129.3	.000
4.8705	2.995	128.9	.000
4.8852	2.995	128.9	.000
4.9000	2.995	129.1	.000
4.9147	2.995	129.1	.000
4.9295	2.987	129.8	.000
4.9443	2.987	129.8	.000
4.9590	2.982	130.2	.000
4.9738	2.982	130.2	.000
4.9885	3.011	129.7	.000

TABLE 6.9   Valves (at node J26-J9-J15-J17-J20-J28) data table Created by Hammer—Version 07.00.049.00 compared to equation of Regression software SPSS.

** Air valve at node J26 **				
Time	Volume	Head	Mass	Air-Flow
(s)	(m3)	(m)	(kg)	(cms)
0.0000	.000	134.99	.0000	.000
4.9885	.000	134.95	.0000	.000

TABLE 6.9 *(Continued)*

** Air valve at node J9 **				
Time	Volume	Head	Mass	Air-Flow
(s)	(m3)	(m)	(kg)	(cms)
0.0000	.000	130.18	.0000	.000
4.9885	.000	129.75	.0000	.000

** Air valve at node J15 **				
Time	Volume	Head	Mass	Air-Flow
(s)	(m3)	(m)	(kg)	(cms)
0.0000	.000	115.22	.0000	.000
4.9885	.000	121.75	.0000	.000

** Air valve at node J17 **				
Time	Volume	Head	Mass	Air-Flow
(s)	(m3)	(m)	(kg)	(cms)
0.0000	.000	113.01	.0000	.000
4.9885	.000	133.60	.0000	.000

** Air valve at node J20 **				
Time	Volume	Head	Mass	Air-Flow
(s)	(m3)	(m)	(kg)	(cms)
0.0000	.000	112.65	.0000	.000
4.9885	.000	118.03	.0000	.000

** Air valve at node J28 **				
Time	Volume	Head	Mass	Air-Flow
(s)	(m3)	(m)	(kg)	(cms)
0.0000	.000	104.55	.0000	.000
4.9885	.000	105.10	.0000	.000

** Surge tank at node J4 **				
Time	Level	Head	Inflow	Spll-Rate
(s)	(m)	(m)	(cms)	(cms)
0.0000	135.0	135.0	.000	.000
4.9885	135.0	135.0	.002	.000

6.3.2 Comparison of Present research results with other expert's research

Comparison of present research results (water hammer software modeling and SPSS modeling), with other expert's research results, shows similarity and advantages:

6.3.2.1 Wylie, E. B., and Streeter, V. L., 1982

Classical water hammer theory neglects convective terms, and assumes fluid wave speed, c (m/s) is dependent on the support conditions of the pipes. Among these methods, MOC-based schemes are most popular because these schemes provide the desirable attributes of accuracy, numerical efficiency and programming simplicity. This method has been used in present research [12].

6.3.2.2 Arris S Tijsseling, Alan E Vardy, 2002

Present study assumed 3 states in field tests; transmission line with surge tank and water hammer in leakage and no leakage condition. Present results have been compared with: Arris S Tijsseling and Alan E Vardy, 2002 .Comparison shows similarity in results [13].

6.3.2.3 Ghidaoui and, Leyn et al., 2005

The efficiency of a model is a critical factor for Real-Time Control (RTC), since several simulations are required within a control loop in order to optimize the control strategy. Small simulation time steps are needed to reproduce the rapidly varying hydraulics. Comparison shows similarity in Researches results [14].

6.3.2.4 Arturo S. Leon, 2007

Comparison shows similarity between present research results with results observed by Arturo S. Leon, 2007 [15].

6.3.2.5 Apoloniusz Kodura, Katarzyna Weinerowska, 2005

Detailed conclusions drawn on the basis of experiments and calculations for the pipeline with a local leak are also presented in the work of Kodura and Weinerowska, 2005 [16-17] (Figure 6.6).

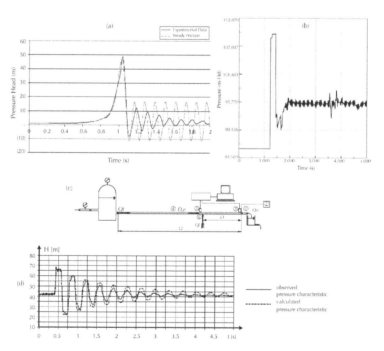

FIGURE 6.6 (a) Pressure Head Histories for a Single Pipe system, Using Steady and unsteady Friction. (Arturo S. Leon Research, 2007). (b) Rasht city Water Pipeline. (c) Pipeline with local leak, (d) Example of the measured and calculated pressure characteristics for the pipeline with local leak

6.4 CONCLUSION

Rapidly closing or opening a valve causes pressure transients in pipelines, known as water hammer. Valve closure can result in pressure well over the steady state values, while valve opening can cause seriously low pressure, possibly so low that the flowing liquid vaporizes inside the pipe. In this study it is assumed that air valves are located at node J26-J9-J15-J17-J20-J28 while leakage location is an opened gate valve (Table 6.2–6.9). Flow changes from 3000 (l/s) down to Min. value of 2,750 (l/s) after 0.0295 (s) and after 4.9885 (s) by harmonic motion reached to 3,011 (l/s). Rate and location of water leakage on water pipeline due to the water hammer pressure has been evaluated by PLC. In order to be able to automatically control the water pipeline and surge tank, GIS has detected rate and Geo reference location of water leakage from PLC. So decreasing of UFW (as the main effect of water hammer disaster) has been showed by practical way in present chapter.

KEYWORDS

- **Euler and continuity equations**
- **Field tests**
- **Geography information system**
- **Method of characteristics**
- **Pressurized system**
- **Programmable logic control**
- **Regression software**
- **Unaccounted for water**

REFERENCES

1. Neshan, H. (1985). Water Hammer, pumps Iran Co. Tehran, Iran, pp. 1-60.
2. Streeter, V. L. and Wylie, E. B. (1979). Fluid Mechanics, McGraw-Hill Ltd., USA, pp. 492-505.
3. Hariri, K (2007a). Decreasing of unaccounted for water "UFW" by Geographic Information System "GIS" in Rasht Urban Water System. Technical and Art. *J. Civil Engineering Organization of Gilan* 38, 3-7.
4. Hariri, K. (2007b). GIS and water hammer disaster at earthquake in Rasht water pipeline. 3rd International Conference on Integrated Natural Disaster Management (INDM) .Tehran, Iran.
5. Hariri, K. (2007c). Interpenetration of two fluids at parallel between plates and turbulent moving in pipe. 8th Conference on Ministry of Energetic works at research week. Tehran, Iran.
6. Hariri, K. (2007d). Water hammer and valves. 8th Conference on Ministry of Energetic works at research week. Tehran, Iran.
7. Hariri, K. (2007e). Water hammer and hydrodynamics' instability. 8th Conference on Ministry of Energetic works at research week. Tehran, Iran.
8. Hariri, K. (2007f). Water hammer analysis and formulation. 8th Conference on Ministry of Energetic works at research week. Tehran, Iran.
9. Hariri, K. (2007g). Water hammer and fluid condition. 8th Conference on Ministry of Energetic works at research week. Tehran, Iran.

10. Hariri, K. (2007h). Water hammer and pump pulsation. 8th Conference on Ministry of Energetic works at research week. Tehran, Iran.

11. Hariri, K. (2007i). Reynolds number and hydrodynamics' instability. 8th Conference on Ministry of Energetic works at research week. Tehran, Iran.

12. Wylie, E. B. and Streeter, V. L. (1982). Fluid Transients, Feb Press, Ann Arbor, MI, 1983. corrected copy: 166-171.

13. Arris S Tijsseling (2004). "Alan E Vardy Time scales and FSI in unsteady liquid-filled pipe flow" 5-12.

14. Ghidaoui and, León et al. (2005). An efficient second-order accurate shock-capturing scheme for modeling one and two-phase water hammer flows. Ph. D. Thesis, 4-44.

15. Arturo, S. Leon (2007) .An efficient second-order accurate shock-capturing scheme for modeling one and two-phase water hammer flows. Ph. D. Thesis, 4-44.

16. Apoloniusz, Kodura, Katarzyna and Weinerowska (2005). Some Aspects of Physical and Numerical Modeling of Water Hammer in Pipelines, pp. 126-132.

17. Kodura, A. and Weinerowska, K. (2005). Some aspects of physical and numerical modeling of water hammer in pipelines. In *International symposium on water management and hydraulic engineering*, pp. 125-33.

7 Applied Hydraulic Transients: Automation and Advanced Control

CONTENTS

NOMENCLATURES

λ = coefficient of combination,

t = time,

$\rho 1$ = density of the light fluid (kg/m3),

$\rho 2$ = density of the heavy fluid (kg/m3),

s = length,

τ = shear stress,

C = surge wave velocity (m/s),

v2-v1 = velocity difference (m/s),

e = pipe thickness (m),

K = module of elasticity of water(kg/m²),

C = wave velocity(m/s),

u = velocity (m/s),

D = Diameter of each pipe (m),

w = weight

$\lambda_.$ = unit of length

V = velocity

C = surge wave velocity in pipe

f = friction factor

H2-H1 = pressure difference (m-H2O)

g = acceleration of gravity (m/s²)

V = volume

Ee = module of elasticity(kg/m²)

θ = Mixed ness integral measure

σ = viscous stress tensor

c = speed of pressure wave (celerity-m/s)

f = Darcy-Weisbach friction factor

θ = Mixed ness integral measure,

R = pipe radius (m²),

J = junction point (m),

A = Pipe cross-sectional area (m²)

d = pipe diameter(m),

Ev = bulk modulus of elasticity,

P = surge pressure (m),

C = Velocity of surge wave (m/s),

ΔV= changes in velocity of water (m/s),

Tp = pipe thickness (m),

Ew = module of elasticity of water (kg/m²),

T = Time (s),

μ = fluid dynamic viscosity(kg/m.s)

γ = specific weight (N/m³)

I = moment of inertia (m^4)

r= pipe radius (m)

dp = is subjected to a static pressure rise

α = kinetic energy correction factor

ρ = density (kg/m3)

g = acceleration of gravity (m/s²)

K = wave number

Ep = pipe module of elasticity (kg/m²)

C1=pipe support coefficient

Ψ = depends on pipeline support-Characteristics and Poisson's ratio

7.1 INTRODUCTION

Analysis, design, and operational procedures all benefit from computer simulations. The study of hydraulic transients is generally considered to have begun with the works of Joukowsky (1898) and Allievi (1902). The historical development of this subject makes for good reading. A number of pioneers made breakthrough contributions to the field, including R. Angus and John Parmakian (1963) and Wood (1970), who popularized and refined the graphical calculation method. Benjamin Wylie and Victor Streeter (1993) combined the method of characteristics (MOC) with computer modeling. The field of fluid transients is still rapidly evolving worldwide by Brunone *et al.*, (2000); Koelle and Luvizotto, (1996); Filion and Karney, (2002); Hamam and McCorquodale, (1982); Savic and Walters, (1995); Walski and Lutes, (1994); Wu and Simpson, (2000). Various methods have been developed to solve transient flow in pipes. This range has been formed from approximate equations to numerical solutions of the nonlinear Navier–Stokes equations. The similarity of the transient conditions caused by different source devices provides the key to transient analysis in a wide range of different systems by understanding the initial state of the system and the ways in which energy and mass are added or removed from it. This is best illustrated by an example for a present research pumping system (Figure 7.1).

This chapter refers to a fluid transient as a "Dynamic" operating case, which may also include sudden thrust due to relief valves that pop open or rapid piping accelerations due to an earthquake. So it is advisable to investigate fluid-structure interpenetration (FSI). Model design need to find the relation between two or many of variables accordance to fluid transient as a "Dynamic" operating [1-3]. In this chapter, transient flow characteristics in pipe have been studied by using new computational techniques. For data collection process, the main pipeline of water networks of Rasht city in the north of Iran have been used (Research Field Tests Model). Model of main pipeline of networks have been selected for the city of Rasht, Guilan province (with 1,050,000

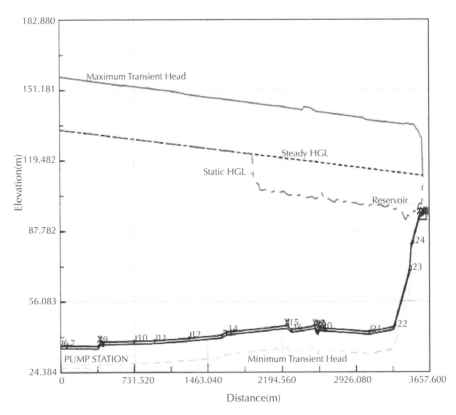

FIGURE 7.1   Typical locations where transient pulses initiate.

population). Data were collected from the "PLC" of water treatment plant pump stations. Water hammer models of the Rasht city pipeline guided route selection, conceptual and detailed design of pipeline. The pipeline is included water treatment plant pump station (in the start of water transmission), 3.595 km of 2*1,200 mm diameter pre-stressed pipes and one 50,000 m³ reservoir (at the end of water transmission). All of these parts have been tied into existing water networks. Long-distance water transmission lines must be economical, reliable and expandable. So this study was retained to provide hydraulic input to a network-wide optimization and risk-reduction strategy for Rasht city main pipeline. Research track record includes multi-booster pressurized lines with surge protection ranging from check valves to gas vessels. This study included particular expertise designing pressurized and pipeline segments, reducing unaccounted for water "UFW" flow where possible to reduce energy costs. Several experiences demonstrated to ensure reliable water transmission for the Rasht city main pipeline. The water hammer phenomena investigation due to fluid condition was the main objective of this research. It formed by investigation of relation between: P-surge pressure(as a function or dependent variable with nomenclature "Y") and several factors(such as independent variables with nomenclature "X") such as:

ρ—density (kg/m³), C–velocity of surge wave (m/s), g—acceleration of gravity (m/s²), ΔV—changes in velocity of water (m/s), d—pipe diameter(m), T—pipe thickness (m), Ep—pipe module of elasticity ((kg/m²), Ew—module of elasticity of water ((kg/m²), C1—pipe support coefficient, T—Time (sec), Tp—pipe thickness (m). Regression software has fitted the function curve and provided regression analysis. Hence the model has been found in the final procedure. By this model, Research Field Tests and lab model results have been compared in the final procedure. This is another practical aim of present chapter. So a Condition Base Maintenance (CM) method has been founded for all water transmission systems [4-6].

7.2   MATERIALS AND METHODS

Present chapter has investigated fluid condition by fluid pressure variations. The interpenetration between the mixture components could be regarded as a combination of the diffusing process and remixing process. If the former dominates the interpenetration, it is miscible; otherwise, it is immiscible. The complex microscopic interplay between the mixture components makes the simulation highly challenging. So far, there have been some dedicated reports in computer graphics dealing with immiscible mixtures, but only few works focused on miscible mixtures. Regression Equations have showed relation between two or many physical units of variables. For example, there was a relation between volume of gases and their internal temperatures. The main approach in this research was investigation of relation between: P—surge pressure (as a function or dependent variable with nomenclature "Y") and several factors (as independent variables with nomenclature "X") such as: ρ—density (kg/m³), C—velocity of surge wave (m/s), g—acceleration of gravity (m/s²), ΔV—changes in velocity of water (m/s), d—pipe diameter (m), T—pipe thickness (m), Ep—pipe module of elasticity ((kg/m²), Ew—module of elasticity of water ((kg/m²), C1—pipe support coefficient, T—Time (sec), Tp—pipe thickness (m), (for fluid at water hammer condition) [7-8].

7.3   RESULTS AND DISCUSSION

(a) Research focus attention on the steam-water flows that may occur during transients formed in pressurized-water pipeline. The method presented is applicable to liquid–solid and liquid–liquid flows as well.

(b) The Laboratory has handled the development of sophisticated numerical techniques for analysis of multiphase flows and in the construction of computer codes based on these techniques (Table 7.1).

(c) When collisions dominate a region with many macro particles, those macro particles can be merged, and their fluid like behavior modeled with a hydrodynamics calculation.

(d) Equations were solved numerically by the MOC. Transient analysis has solved transient flow in pipes for range of approximate equations to numerical solutions of the nonlinear Navier–Stokes equations.

7.3.1 Governing Equations for Unsteady (or Transient) Flow

Hydraulic transient flow is also known as unsteady fluid flow. During a transient analysis, the fluid and system boundaries can be either elastic or inelastic:

- **Elastic theory** describes unsteady flow of a compressible liquid in an elastic system (e.g., where pipes can expand and contract). Research uses the MOC to solve virtually any hydraulic transient problems.

- **Rigid-column theory** describes unsteady flow of an incompressible liquid in a rigid system. It is only applicable to slower transient phenomena. Both branches of transient theory stem from the same governing equations. Transient analysis results that are not comparable with actual system measurements are generally caused by inappropriate system data (especially boundary conditions) and inappropriate assumptions [9-11].

TABLE 7.1 Input data table including of -"x" & "x²" & "y"-for equation finding, transfer to Regression software "SPSS".

$$y = a_0 + a_1 x \qquad \text{(Line Equations)}$$

Equation	Model Summary					Parameter Estimates			
	R Square	F	df1	df2	Sig.	Constant	b1	b2	b3
Linear	.418	15.831	1	22	.001	6.062	.571		

$$y = a_0 + a_1 x + a_2 x^2 \qquad \text{(Second order Equation)}$$

Equation	Model Summary					Parameter Estimates			
	R Square	F	df1	df2	Sig.	Constant	b1	b2	b3
Quadratic	.487	9.955	2	21	.001	6.216	-.365	.468	

$$y = a_0 + a_1 x + a_2 x^2 + a_3 x^3 \qquad \text{(Third order Equation)}$$

Equation	Model Summary					Parameter Estimates			
	R Square	F	df1	df2	Sig.	Constant	b1	b2	b3
Cubic	.493	10.193	2	21	.001	6.239	.000	-.057	.174

$$A = Ce^{kt} \qquad \text{(Compound)}$$

Equation	Model Summary					Parameter Estimates			
	R Square	F	df1	df2	Sig.	Constant	b1	b2	b3
Compound	.424	16.207	1	22	.001	6.076	1.089		

$$(dA/dT) = KA, \ \text{(Growth)}$$

Equation	Model Summary					Parameter Estimates			
	R Square	F	df1	df2	Sig.	Constant	b1	b2	b3
Growth	.424	16.207	1	22	.001	1.804	.085		

$$y = ab^x \ or \quad \log a + x \log b = a_0 + a_1 x \quad \text{(Exponential Equation)}$$

$$y = ab^x + g \qquad \text{(Expression Exponential Equation)}$$

Equation	Model Summary					Parameter Estimates			
	R Square	F	df1	df2	Sig.	Constant	b1	b2	b3
Exponential	.424	16.207	1	22	.001	6.076	.085		

$$y = ax^b \quad or \quad \log y = \log a + b \log x \qquad \text{(Logarithmic Equation)}$$

$$y = ax^b + g \qquad \text{(Expression Logarithmic Equation)}$$

Equation	Model Summary					Parameter Estimates			
	R Square	F	df1	df2	Sig.	Constant	b1	b2	b3
Logistic	.424	16.207	1	22	.001	.165	.918		

Research Field Tests Model (water pipeline of Rasht city in the north of Iran)
Software Hammer—Version 07.00.049.00
Type of Run: Full
Date of Run: 09/19/08
Time of Run: 04:47 am
Data File: E:\k-hariri Asli\daraye nashti.inp
Hydrograph File: Not Selected
Labels: Short

TABLE 7.2   (a) System information preference, (b) Debug information.

(a)	
ITEM	VALUE
Time Step (s)	
Automatic	0.0148
User-Selected	n/a
Total Number	339
Total Simulation Time (s)	5.00
Units	cms, m
Total Number of Nodes	27
Total Number of Pipes	26

TABLE 7.2 *(Continued)*

Specific Gravity	1.00
Wave Speed (m/s)	1084
Vapor Pressure (m)	-10.0
Reports	
Number of Nodes	52
Number of Time Steps	All
Number of Paths	1
Output	
Standard	No
Cavities (Open/Close)	No
Adjustments	
Adjusted Variable	Length
Warning Limit (%)	75.00
Calculate Transient Forces	No
Use Auxiliary Data File	Yes

(b)

ITEM	VALUE
Tolerances	
Initial Flow Consistency Value	0.0006
Initial Head Consistency Value	0.030
Criterion for Fr. Coef. Flag	0.025
Adjustments	
Elevation Decrease	0.00
Extreme Heads Display	
All Times	Yes
After First Extreme	No
Friction Coefficient	
Model	Steady
1,000,000 x Kinematic Viscosity	n/a
Debug Parameters	
Level	Null

TABLE 7.3 (a) The total paths, (b) The paths in 26 points, table created by Hammer—Version 07.00.049.00 compared to equation of Regression software "SPSS".

(a)

Path	No. of Points	From Point	To Point	Length (m)
1	249	P2:J6	P25:N1	3595.0
P2	5	P2:J6	P2:J7	60.7

TABLE 7.3 *(Continued)*

			(b)	
Pipe	No. of Points	From Point	To Point	Length (m)
P3	21	P3:J7	P3:J8	311.0
P4	2	P4:J3	P4:J4	1.0
P5	2	P5:J4	P5:J26	.5
P6	8	P6:J26	P6:J27	108.7
P7	3	P7:J27	P7:J6	21.5
P8	2	P8:J8	P8:J9	15.0
P9	23	P9:J9	P9:J10	340.7
P10	14	P10:J10	P10:J11	207.0
P11	22	P11:J11	P11:J12	339.0
P12	22	P12:J12	P12:J13	328.6
P13	4	P13:J13	P13:J14	47.0
P14	38	P14:J14	P14:J15	590.0
P15	4	P15:J15	P15:J16	49.0
P16	15	P16:J16	P16:J17	224.0
P17	2	P17:J17	P17:J18	18.4
P18	2	P18:J18	P18:J19	14.6
P19	2	P19:J19	P19:J20	12.0
P20	32	P20:J20	P20:J21	499.0
P21	17	P21:J21	P21:J22	243.4
P22	11	P22:J22	P22:J23	156.0
P23	3	P23:J23	P23:J24	22.0
P24	6	P24:J24	P24:J28	82.0
P25	4	P25:J28	P25:N1	35.6
P0	2	P0:J1	P0:J2	.5
P1	2	P1:J2	P1:J3	.5

TABLE 7.4 Pipe information created by Hammer—Version 07.00.049.00 (Snapshot for selected end points at time 4.9885 s) compared to equation of Regression software.

LENGTH LABEL	DIAMETER (m)	WAVESPEED (mm)	D-W (m/s)	FR. COEF.	CHECK VALVE
P2	60.7	1200.00	1084.0	0.028	*
P3	311.0	1200.00	1084.0	0.028	*
P4	1.0	1200.00	1084.0	0.027	*
P5	0.5	1200.00	1084.0	0.027	*
P6	108.7	1200.00	1084.0	0.067	*
P7	21.5	1200.00	1084.0	0.028	*

TABLE 7.4 *(Continued)*

LENGTH LABEL	DIAMETER (m)	WAVESPEED (mm)	D-W (m/s)	FR. COEF.	VALVE	CHECK
P8	15.0	1200.00	1084.0	0.028	*	
P9	340.7	1200.00	1084.0	0.028	*	
P10	207.0	1200.00	1084.0	0.028	*	
P11	339.0	1200.00	1084.0	0.028	*	
P12	328.6	1200.00	1084.0	0.028	*	
P13	47.0	1200.00	1084.0	0.028	*	
P14	590.0	1200.00	1084.0	0.028	*	
P15	49.0	1200.00	1084.0	0.028	*	
P16	224.0	1200.00	1084.0	0.028	*	
P17	18.4	1200.00	1084.0	0.028	*	
P18	14.6	1200.00	1084.0	0.028	*	
P19	12.0	1200.00	1084.0	0.028	*	
P20	499.0	1200.00	1084.0	0.028	*	
P21	243.4	1200.00	1084.0	0.028	*	
P22	156.0	1200.00	1084.0	0.028	*	
P23	22.0	1200.00	1084.0	0.036	*	
P24	82.0	1200.00	1084.0	0.028	*	
P25	35.6	1200.00	1084.0	0.028	*	
P0	0.5	1200.00	1084.0	0.027	*	
P1	0.5	1200.00	1084.0	0.027	*	

* Darcy-Weisbach friction coefficient exceeds .025.

TABLE 7.5 Initial pipe conditions information created by Hammer—Version 07.00.049.00 (Snapshot for selected end points at time 4.9885 s) compared to equation of Regression software "SPSS".

FROM NODE	HEAD (m)	TO NODE	HEAD (m)	FLOW (cms)	VEL (m/s)
J6	133.4	J7	133.0	2.500	2.21
J7	133.0	J8	131.2	2.500	2.21
J3	135.0	J4	135.0	3.000	2.65
J4	135.0	J26	135.0	3.000	2.65
J26	135.0	J27	133.5	2.500	2.21
J27	133.5	J6	133.4	2.500	2.21
J8	131.2	J9	131.1	2.500	2.21
J9	131.1	J10	129.2	2.500	2.21
J10	129.2	J11	128.0	2.500	2.21

TABLE 7.5   *(Continued)*

FROM NODE	HEAD (m)	TO NODE	HEAD (m)	FLOW (cms)	VEL (m/s)
J11	128.0	J12	126.0	2.500	2.21
J12	126.0	J13	124.1	2.500	2.21
J13	124.1	J14	123.8	2.500	2.21
J14	123.9	J15	120.4	2.500	2.21
J15	120.4	J16	120.2	2.500	2.21
J16	120.2	J17	118.9	2.500	2.21
J17	118.9	J18	118.8	2.500	2.21
J18	118.8	J19	118.7	2.500	2.21
J19	118.7	J20	118.6	2.500	2.21
J20	118.6	J21	115.7	2.500	2.21
J21	115.7	J22	114.3	2.500	2.21
J22	114.3	J23	113.4	2.500	2.21
J23	113.5	J24	113.3	2.500	2.21
J24	113.3	J28	112.8	2.500	2.21
J28	112.8	N1	112.6	2.500	2.21
J1	40.6	J2	40.6	3.000	2.65
J2	40.6	J3	40.6	3.000	2.65

TABLE 7.6   Output data table created by Hammer—Version 07.00.049.00 Min. and Max. head compared to equation of Regression software "SPSS".

END POINT	MAX. PRESS. (mH)	MIN. PRESS. (mH)	MAX. HEAD (m)	MIN. HEAD (m)
P2:J6	124.1	97.0	160.4	133.4
P2:J7	123.7	96.7	160.0	133.0
P3:J7	123.7	96.7	160.0	133.0
P3:J8	121.9	95.1	158.0	131.2
P4:J3	126.7	99.5	162.2	135.0
P4:J4	125.1	97.8	162.3	135.0
P5:J4	125.1	97.8	162.3	135.0
P5:J26	125.1	97.8	162.3	135.0
P6:J26	125.1	97.8	162.3	135.0
P6:J27	124.1	97.0	160.5	133.5
P7:J27	124.1	97.0	160.5	133.5
P7:J6	124.1	97.0	160.4	133.4
P8:J8	121.9	95.1	158.0	131.2
P8:J9	119.9	93.1	157.9	131.1
P9:J9	119.9	93.1	157.9	131.1

TABLE 7.6   *(Continued)*

END POINT	MAX. PRESS. (mH)	MIN. PRESS. (mH)	MAX. HEAD (m)	MIN. HEAD (m)
P9:J10	117.4	90.8	155.7	129.2
P10:J10	117.4	90.8	155.7	129.2
P10:J11	115.8	89.4	154.4	128.0
P11:J11	115.8	89.4	154.4	128.0
P11:J12	112.6	86.4	152.2	126.0
P12:J12	112.6	86.4	152.2	126.0
P12:J13	109.1	83.1	150.1	124.1
P13:J13	109.1	83.1	150.1	124.1
P13:J14	107.5	81.5	149.8	123.8
P14:J14	107.5	81.5	149.8	123.9
P14:J15	100.9	60.9	146.1	106.1
P15:J15	100.9	60.9	146.1	106.1
P15:J16	102.3	62.3	145.8	105.8
P16:J16	102.3	62.3	145.8	105.8
P16:J17	99.3	59.2	144.3	104.3
P17:J17	99.3	59.2	144.3	104.3
P17:J18	101.2	61.1	144.2	104.1
P18:J18	101.2	61.1	144.2	104.1
P18:J19	101.8	61.7	144.1	104.0
P19:J19	101.8	61.7	144.1	104.0
P19:J20	99.8	59.8	144.0	104.0
P20:J20	99.8	59.8	144.0	104.0
P20:J21	98.3	58.1	140.9	100.6
P21:J21	98.3	58.1	140.9	100.6
P21:J22	94.7	54.4	139.3	99.0
P22:J22	94.7	54.4	139.3	99.0
P22:J23	68.4	28.1	138.3	98.0
P23:J23	68.4	28.1	138.3	98.0
P23:J24	56.3	15.9	138.1	97.7
P24:J24	56.3	15.9	138.1	97.7
P24:J28	42.4	0.0	137.6	95.2
P25:J28	42.4	0.0	137.6	95.2
P25:N1	16.7	16.7	112.6	112.6
P0:J1	0.0	0.0	40.6	40.6
P0:J2	27.5	2.1	63.0	37.6
P1:J2	27.5	2.1	63.0	37.6
P1:J3	27.5	2.1	63.0	37.6

Elapsed time: 12 s.

TABLE 7.7   Pipes extreme heads table created by Hammer—Version 07.00.049.00 compared to equation of Regression software "SPSS".

	Distance	Elevation	Init Head	Max Head	Min Head	Max Vol 3	Vap Press
Point	**(m)**	**(m)**	**(m)**	**(m)**	**(m)**	**(m3)**	**(m)**
+ P2:J6	.0	36.3	133.3	157.2	106.0	.000	-10.0
P2:25.00%	15.2	36.3	133.2	157.1	105.9	.000	-10.0
P2:50.00%	30.4	36.3	133.1	157.0	105.8	.000	-10.0
P2:75.00%	45.5	36.3	132.9	156.9	105.7	.000	-10.0
P2:J7	60.7	36.3	132.8	156.8	105.1	.000	-10.0
+ P3:J7	.0	36.3	132.8	156.8	105.1	.000	-10.0
P3:5.00%	15.6	36.3	132.7	156.6	105.3	.000	-10.0
P3:10.00%	31.1	36.3	132.6	156.3	105.1	.000	-10.0
P3:15.00%	46.7	36.3	132.4	156.1	105.0	.000	-10.0
P3:20.00%	62.2	36.3	132.3	155.8	104.9	.000	-10.0
P3:25.00%	77.8	36.3	132.2	155.6	104.7	.000	-10.0
P3:30.00%	93.3	36.2	132.1	155.5	104.4	.000	-10.0
P3:35.00%	108.9	36.2	131.9	155.3	104.3	.000	-10.0
P3:40.00%	124.4	36.2	131.8	155.2	104.2	.000	-10.0
P3:45.00%	140.0	36.2	131.7	155.0	104.0	.000	-10.0
P3:50.00%	155.5	36.2	131.6	135.1	103.8	.000	-10.0
P3:55.00%	171.1	36.2	131.4	134.9	103.7	.000	-10.0
P3:60.00%	186.6	36.2	131.3	134.7	103.5	.000	-10.0
P3:65.00%	202.2	36.2	131.2	134.4	103.4	.000	-10.0
P3:70.00%	217.7	36.2	131.1	134.6	102.9	.000	-10.0
P3:75.00%	233.3	36.2	130.9	134.4	102.7	.000	-10.0
P3:80.00%	248.8	36.1	130.8	134.0	102.7	.000	-10.0
P3:85.00%	264.4	36.1	130.7	133.9	102.4	.000	-10.0
P3:90.00%	279.9	36.1	130.6	133.6	102.3	.000	-10.0
P3:95.00%	295.5	36.1	130.4	133.5	102.1	.000	-10.0
P3:J8	311.0	36.1	130.3	133.3	102.0	.000	-10.0
+ P4:J3	.0	35.5	135.0	135.0	135.0	.000	-10.0
P4:J4	1.0	37.2	135.0	135.0	135.0	.000	-10.0
+ P5:J4	.0	37.2	135.0	135.0	135.0	.000	-10.0
P5:J26	.5	37.2	135.0	136.4	133.7	.000	-10.0
+ P6:J26	.0	37.2	135.0	136.4	133.7	.000	-10.0
P6:14.29%	15.5	37.1	134.8	157.6	111.8	.000	-10.0
P6:28.57%	31.1	37.0	134.6	157.4	109.4	.000	-10.0

TABLE 7.7   *(Continued)*

Point	Distance (m)	Elevation (m)	Init Head (m)	Max Head (m)	Min Head (m)	Max Vol 3 (m3)	Vap Press (m)
P6:42.86%	46.6	36.9	134.3	158.0	108.3	.000	-10.0
P6:57.14%	62.1	36.8	134.1	157.8	107.6	.000	-10.0
P6:71.43%	77.6	36.7	133.9	157.6	106.7	.000	-10.0
P6:85.71%	93.2	36.6	133.7	157.4	106.3	.000	-10.0
P6:J27	108.7	36.5	133.5	157.3	106.7	.000	-10.0
+ P7:J27	.0	36.5	133.5	157.3	106.7	.000	-10.0
P7:50.00%	10.8	36.4	133.4	157.3	106.8	.000	-10.0
P7:J6	21.5	36.3	133.3	157.2	106.0	.000	-10.0
+ P8:J8	.0	36.1	130.3	133.3	102.0	.000	-10.0
P8:J9	15.0	38.0	130.2	133.2	101.8	.000	-10.0
+ P9:J9	.0	38.0	130.2	133.2	101.8	.000	-10.0
P9:4.55%	15.5	38.0	130.1	133.1	101.7	.000	-10.0
P9:9.09%	31.0	38.0	129.9	133.0	101.9	.000	-10.0
P9:13.64%	46.5	38.0	129.8	132.8	101.8	.000	-10.0
P9:18.18%	61.9	38.1	129.7	132.7	101.7	.000	-10.0
P9:22.73%	77.4	38.1	129.6	132.4	101.4	.000	-10.0
P9:27.27%	92.9	38.1	129.4	132.3	101.3	.000	-10.0
P9:31.82%	108.4	38.1	129.3	132.2	101.1	.000	-10.0
P9:36.36%	123.9	38.1	129.2	132.0	101.1	.000	-10.0
P9:40.91%	139.4	38.1	129.1	131.9	101.0	.000	-10.0
P9:45.45%	154.9	38.1	128.9	131.7	100.9	.000	-10.0
P9:50.00%	170.4	38.2	128.8	131.7	100.8	.000	-10.0
P9:54.55%	185.8	38.2	128.7	131.5	100.6	.000	-10.0
P9:59.09%	201.3	38.2	128.6	131.3	100.5	.000	-10.0
P9:63.64%	216.8	38.2	128.4	131.2	100.4	.000	-10.0
P9:68.18%	232.3	38.2	128.3	131.5	100.2	.000	-10.0
P9:72.73%	247.8	38.2	128.2	131.7	100.1	.000	-10.0
P9:77.27%	263.3	38.2	128.1	131.5	100.0	.000	-10.0
P9:81.82%	278.8	38.3	127.9	131.3	99.8	.000	-10.0
P9:86.36%	294.2	38.3	127.8	131.2	99.7	.000	-10.0
P9:90.91%	309.7	38.3	127.7	131.0	99.5	.000	-10.0
P9:95.45%	325.2	38.3	127.6	130.9	99.5	.000	-10.0
P9:J10	340.7	38.3	127.4	130.8	99.3	.000	-10.0
+ P10:J10	.0	38.3	127.4	130.8	99.3	.000	-10.0
P10:7.69%	15.9	38.3	127.3	130.7	99.2	.000	-10.0
P10:15.38%	31.8	38.4	127.2	130.6	99.0	.000	-10.0

TABLE 7.7 *(Continued)*

	Distance	Elevation	Init Head	Max Head	Min Head	Max Vol 3	Vap Press
Point	(m)	(m)	(m)	(m)	(m)	(m3)	(m)
P10:23.08%	47.8	38.4	127.0	130.4	98.9	.000	-10.0
P10:30.77%	63.7	38.4	126.9	129.9	99.2	.000	-10.0
P10:38.46%	79.6	38.4	126.8	129.8	99.1	.000	-10.0
P10:46.15%	95.5	38.4	126.7	129.8	98.8	.000	-10.0
P10:53.85%	111.5	38.5	126.5	129.7	98.7	.000	-10.0
P10:61.54%	127.4	38.5	126.4	129.6	98.6	.000	-10.0
P10:69.23%	143.3	38.5	126.3	129.5	98.4	.000	-10.0
P10:76.92%	159.2	38.5	126.1	129.3	98.3	.000	-10.0
P10:84.62%	175.2	38.5	126.0	129.2	98.2	.000	-10.0
P10:92.31%	191.1	38.6	125.9	129.1	98.1	.000	-10.0
P10:J11	207.0	38.6	125.8	129.0	97.9	.000	-10.0
+ P11:J11	.0	38.6	125.8	129.0	97.9	.000	-10.0
P11:4.76%	16.1	38.6	125.6	128.8	97.5	.000	-10.0
P11:9.52%	32.3	38.7	125.5	128.7	97.3	.000	-10.0
P11:14.29%	48.4	38.7	125.4	128.5	97.2	.000	-10.0
P11:19.05%	64.6	38.8	125.2	128.7	97.1	.000	-10.0
P11:23.81%	80.7	38.8	125.1	128.4	97.1	.000	-10.0
P11:28.57%	96.9	38.9	125.0	128.2	96.9	.000	-10.0
P11:33.33%	113.0	38.9	124.8	128.2	96.8	.000	-10.0
P11:38.10%	129.1	39.0	124.7	128.0	96.6	.000	-10.0
P11:42.86%	145.3	39.0	124.6	127.9	96.5	.000	-10.0
P11:47.62%	161.4	39.1	124.5	127.8	96.4	.000	-10.0
P11:52.38%	177.6	39.1	124.3	127.7	96.3	.000	-10.0
P11:57.14%	193.7	39.2	124.2	127.6	96.1	.000	-10.0
P11:61.90%	209.9	39.2	124.1	127.4	96.0	.000	-10.0
P11:66.67%	226.0	39.3	123.9	127.0	96.2	.000	-10.0
P11:71.43%	242.1	39.3	123.8	127.0	95.9	.000	-10.0
P11:76.19%	258.3	39.4	123.7	126.9	95.7	.000	-10.0
P11:80.95%	274.4	39.4	123.5	126.7	95.7	.000	-10.0
P11:85.71%	290.6	39.5	123.4	126.7	95.6	.000	-10.0
P11:90.48%	306.7	39.5	123.3	126.5	95.5	.000	-10.0
P11:95.24%	322.9	39.6	123.1	126.4	95.3	.000	-10.0
P11:J12	339.0	39.7	123.0	126.1	95.3	.000	-10.0
+ P12:J12	.0	39.7	123.0	126.1	95.3	.000	-10.0
P12:4.76%	15.6	39.7	122.9	126.1	95.2	.000	-10.0
P12:9.52%	31.3	39.8	122.8	125.9	95.1	.000	-10.0

TABLE 7.7   *(Continued)*

Point	Distance (m)	Elevation (m)	Init Head (m)	Max Head (m)	Min Head (m)	Max Vol 3 (m3)	Vap Press (m)
P12:14.29%	46.9	39.8	122.6	125.7	95.0	.000	-10.0
P12:19.05%	62.6	39.9	122.5	125.6	95.0	.000	-10.0
P12:23.81%	78.2	40.0	122.4	125.5	94.9	.000	-10.0
P12:28.57%	93.9	40.0	122.3	125.4	94.8	.000	-10.0
P12:33.33%	109.5	40.1	122.1	125.1	94.7	.000	-10.0
P12:38.10%	125.2	40.2	122.0	125.0	94.5	.000	-10.0
P12:42.86%	140.8	40.2	121.9	130.6	94.4	.000	-10.0
P12:47.62%	156.5	40.3	121.8	130.4	94.4	.000	-10.0
P12:52.38%	172.1	40.4	121.6	130.6	94.2	.000	-10.0
P12:57.14%	187.8	40.4	121.5	130.6	94.1	.000	-10.0
P12:61.90%	203.4	40.5	121.4	132.1	94.0	.000	-10.0
P12:66.67%	219.1	40.5	121.2	132.0	93.9	.000	-10.0
P12:71.43%	234.7	40.6	121.1	131.9	93.8	.000	-10.0
P12:76.19%	250.4	40.7	121.0	131.8	93.7	.000	-10.0
P12:80.95%	266.0	40.7	120.9	131.7	93.4	.000	-10.0
P12:85.71%	281.7	40.8	120.7	131.5	93.3	.000	-10.0
P12:90.48%	297.3	40.9	120.6	133.4	93.2	.000	-10.0
P12:95.24%	313.0	40.9	120.5	133.3	93.0	.000	-10.0
P12:J13	328.6	41.0	120.4	133.1	92.9	.000	-10.0
+ P13:J13	.0	41.0	120.4	133.1	92.9	.000	-10.0
P13:33.33%	15.7	41.4	120.2	133.0	92.8	.000	-10.0
P13:66.67%	31.3	41.9	120.1	132.9	92.7	.000	-10.0
P13:J14	47.0	42.3	120.0	132.7	92.5	.000	-10.0
+ P14:J14	.0	42.3	120.0	132.7	92.5	.000	-10.0
P14:2.70%	15.9	42.4	119.9	132.6	92.4	.000	-10.0
P14:5.41%	31.9	42.5	119.7	141.9	92.3	.000	-10.0
P14:8.11%	47.8	42.6	119.6	150.0	92.1	.000	-10.0
P14:10.81%	63.8	42.6	119.5	150.1	92.0	.000	-10.0
P14:13.51%	79.7	42.7	119.3	150.0	91.9	.000	-10.0
P14:16.22%	95.7	42.8	119.2	149.9	91.8	.000	-10.0
P14:18.92%	111.6	42.9	119.1	149.8	91.4	.000	-10.0
P14:21.62%	127.6	42.9	119.0	149.5	91.3	.000	-10.0
P14:24.32%	143.5	43.0	118.8	149.4	91.2	.000	-10.0
P14:27.03%	159.5	43.1	118.7	149.3	91.3	.000	-10.0
P14:29.73%	175.4	43.2	118.6	149.3	91.2	.000	-10.0
P14:32.43%	191.4	43.3	118.4	149.2	91.0	.000	-10.0

TABLE 7.7 *(Continued)*

Point	Distance (m)	Elevation (m)	Init Head (m)	Max Head (m)	Min Head (m)	Max Vol 3 (m3)	Vap Press (m)
P14:35.14%	207.3	43.3	118.3	149.1	90.9	.000	-10.0
P14:37.84%	223.2	43.4	118.2	148.9	90.8	.000	-10.0
P14:40.54%	239.2	43.5	118.0	148.9	90.7	.000	-10.0
P14:43.24%	255.1	43.6	117.9	148.7	90.6	.000	-10.0
P14:45.95%	271.1	43.6	117.8	148.4	90.7	.000	-10.0
P14:48.65%	287.0	43.7	117.7	148.6	90.6	.000	-10.0
P14:51.35%	303.0	43.8	117.5	148.5	90.5	.000	-10.0
P14:54.05%	318.9	43.9	117.4	148.4	90.1	.000	-10.0
P14:56.76%	334.9	44.0	117.3	148.2	90.0	.000	-10.0
P14:59.46%	350.8	44.0	117.1	148.2	89.9	.000	-10.0
P14:62.16%	366.8	44.1	117.0	148.1	89.1	.000	-10.0
P14:64.86%	382.7	44.2	116.9	148.0	88.2	.000	-10.0
P14:67.57%	398.6	44.3	116.8	147.7	90.6	.000	-10.0
P14:70.27%	414.6	44.3	116.6	147.5	90.3	.000	-10.0
P14:72.97%	430.5	44.4	116.5	147.6	89.3	.000	-10.0
P14:75.68%	446.5	44.5	116.4	147.5	89.2	.000	-10.0
P14:78.38%	462.4	44.6	116.2	147.4	89.0	.000	-10.0
P14:81.08%	478.4	44.7	116.1	147.3	88.7	.000	-10.0
P14:83.78%	494.3	44.7	116.0	146.9	88.6	.000	-10.0
P14:86.49%	510.3	44.8	115.9	146.7	88.5	.000	-10.0
P14:89.19%	526.2	44.9	115.7	146.5	88.5	.000	-10.0
P14:91.89%	542.2	45.0	115.6	146.5	88.0	.000	-10.0
P14:94.59%	558.1	45.1	115.5	146.4	86.0	.000	-10.0
P14:97.30%	574.1	45.1	115.3	146.3	85.1	.000	-10.0
P14:J15	590.0	45.2	115.2	146.2	88.0	.000	-10.0
+ P15:J15	.0	45.2	115.2	146.2	88.0	.000	-10.0
P15:33.33%	16.3	44.6	115.1	146.2	87.6	.000	-10.0
P15:66.67%	32.7	44.0	115.0	146.2	88.0	.000	-10.0
P15:J16	49.0	43.4	114.8	146.0	87.9	.000	-10.0
+ P16:J16	.0	43.4	114.8	146.0	87.9	.000	-10.0
P16:7.14%	16.0	43.6	114.7	146.2	87.5	.000	-10.0
P16:14.29%	32.0	43.7	114.6	146.3	87.0	.000	-10.0
P16:21.43%	48.0	43.8	114.4	146.1	87.4	.000	-10.0
P16:28.57%	64.0	43.9	114.3	146.0	87.3	.000	-10.0
P16:35.71%	80.0	44.0	114.2	145.8	87.2	.000	-10.0
P16:42.86%	96.0	44.1	114.0	145.8	86.5	.000	-10.0

TABLE 7.7   *(Continued)*

Point	Distance (m)	Elevation (m)	Init Head (m)	Max Head (m)	Min Head (m)	Max Vol 3 (m3)	Vap Press (m)
P16:50.00%	112.0	44.2	113.9	145.7	85.6	.000	-10.0
P16:57.14%	128.0	44.3	113.8	145.5	87.8	.000	-10.0
P16:64.29%	144.0	44.5	113.7	147.5	86.9	.000	-10.0
P16:71.43%	160.0	44.6	113.5	147.4	85.9	.000	-10.0
P16:78.57%	176.0	44.7	113.4	147.2	86.7	.000	-10.0
P16:85.71%	192.0	44.8	113.3	147.2	86.9	.000	-10.0
P16:92.86%	208.0	44.9	113.1	147.2	86.1	.000	-10.0
P16:J17	224.0	45.0	113.0	146.9	85.9	.000	-10.0
+ P17:J17	.0	45.0	113.0	146.9	85.9	.000	-10.0
P17:J18	18.4	43.0	112.9	147.3	85.9	.000	-10.0
+ P18:J18	.0	43.0	112.9	147.3	85.9	.000	-10.0
P18:J19	14.6	42.3	112.7	146.2	86.5	.000	-10.0
+ P19:J19	.0	42.3	112.7	146.2	86.5	.000	-10.0
P19:J20	12.0	44.2	112.6	145.9	85.8	.000	-10.0
+ P20:J20	.0	44.2	112.6	145.9	85.8	.000	-10.0
P20:3.23%	16.1	44.2	112.5	145.2	86.3	.000	-10.0
P20:6.45%	32.2	44.1	112.4	145.0	86.1	.000	-10.0
P20:9.68%	48.3	44.0	112.3	146.0	86.0	.000	-10.0
P20:12.90%	64.4	44.0	112.1	145.4	85.9	.000	-10.0
P20:16.13%	80.5	43.9	112.0	146.1	85.8	.000	-10.0
P20:19.35%	96.6	43.9	111.9	145.7	85.7	.000	-10.0
P20:22.58%	112.7	43.8	111.7	147.1	85.6	.000	-10.0
P20:25.81%	128.8	43.8	111.6	146.2	85.5	.000	-10.0
P20:29.03%	144.9	43.7	111.5	144.3	85.3	.000	-10.0
P20:32.26%	161.0	43.7	111.3	145.7	85.2	.000	-10.0
P20:35.48%	177.1	43.6	111.2	146.2	84.5	.000	-10.0
P20:38.71%	193.2	43.5	111.1	146.3	84.4	.000	-10.0
P20:41.94%	209.3	43.5	111.0	147.1	84.2	.000	-10.0
P20:45.16%	225.4	43.4	110.8	146.2	84.1	.000	-10.0
P20:48.39%	241.5	43.4	110.7	144.1	84.1	.000	-10.0
P20:51.61%	257.5	43.3	110.6	144.6	84.0	.000	-10.0
P20:54.84%	273.6	43.3	110.4	144.0	83.5	.000	-10.0
P20:58.06%	289.7	43.2	110.3	144.6	83.4	.000	-10.0
P20:61.29%	305.8	43.2	110.2	148.2	83.2	.000	-10.0
P20:64.52%	321.9	43.1	110.0	146.6	83.1	.000	-10.0
P20:67.74%	338.0	43.1	109.9	145.7	83.0	.000	-10.0

TABLE 7.7 *(Continued)*

	Distance	Elevation	Init Head	Max Head	Min Head	Max Vol 3	Vap Press
Point	(m)	(m)	(m)	(m)	(m)	(m3)	(m)
P20:70.97%	354.1	43.0	109.8	144.0	82.9	.000	-10.0
P20:74.19%	370.2	42.9	109.7	141.7	82.8	.000	-10.0
P20:77.42%	386.3	42.9	109.5	144.4	83.2	.000	-10.0
P20:80.65%	402.4	42.8	109.4	144.6	83.0	.000	-10.0
P20:83.87%	418.5	42.8	109.3	143.8	82.9	.000	-10.0
P20:87.10%	434.6	42.7	109.1	142.8	85.4	.000	-10.0
P20:90.32%	450.7	42.7	109.0	144.0	83.6	.000	-10.0
P20:93.55%	466.8	42.6	108.9	144.0	82.7	.000	-10.0
P20:96.77%	482.9	42.6	108.7	144.3	82.5	.000	-10.0
P20:J21	499.0	42.5	108.6	143.1	82.2	.000	-10.0
+ P21:J21	.0	42.5	108.6	143.1	82.2	.000	-10.0
P21:6.25%	15.2	42.6	108.5	144.3	82.1	.000	-10.0
P21:12.50%	30.4	42.8	108.4	144.1	82.1	.000	-10.0
P21:18.75%	45.6	42.9	108.2	142.8	85.0	.000	-10.0
P21:25.00%	60.8	43.0	108.1	142.3	84.9	.000	-10.0
P21:31.25%	76.1	43.2	108.0	142.6	85.1	.000	-10.0
P21:37.50%	91.3	43.3	107.9	142.7	85.0	.000	-10.0
P21:43.75%	106.5	43.4	107.8	142.0	85.5	.000	-10.0
P21:50.00%	121.7	43.5	107.6	141.0	85.4	.000	-10.0
P21:56.25%	136.9	43.7	107.5	141.6	85.3	.000	-10.0
P21:62.50%	152.1	43.8	107.4	141.1	85.2	.000	-10.0
P21:68.75%	167.3	43.9	107.3	141.2	85.1	.000	-10.0
P21:75.00%	182.5	44.1	107.1	140.8	84.9	.000	-10.0
P21:81.25%	197.8	44.2	107.0	141.1	84.8	.000	-10.0
P21:87.50%	213.0	44.3	106.9	140.8	82.1	.000	-10.0
P21:93.75%	228.2	44.5	106.8	140.8	80.3	.000	-10.0
P21:J22	243.4	44.6	106.6	140.2	79.5	.000	-10.0
+ P22:J22	.0	44.6	106.7	140.2	79.5	.000	-10.0
P22:10.00%	15.6	47.1	106.5	141.6	80.3	.000	-10.0
P22:20.00%	31.2	49.7	106.4	141.0	79.5	.000	-10.0
P22:30.00%	46.8	52.2	106.3	141.1	83.2	.000	-10.0
P22:40.00%	62.4	54.7	106.1	140.7	81.4	.000	-10.0
P22:50.00%	78.0	57.2	106.0	141.9	79.6	.000	-10.0
P22:60.00%	93.6	59.8	105.9	141.8	76.4	.000	-10.0
P22:70.00%	109.2	62.3	105.8	142.2	74.6	.000	-10.0
P22:80.00%	124.8	64.8	105.6	144.1	73.8	.000	-10.0

TABLE 7.7 *(Continued)*

Point	Distance (m)	Elevation (m)	Init Head (m)	Max Head (m)	Min Head (m)	Max Vol 3 (m3)	Vap Press (m)
P22:90.00%	140.4	67.4	105.5	144.2	77.2	.000	-10.0
P22:J23	156.0	69.9	105.4	140.4	76.4	.000	-10.0
+ P23:J23	.0	69.9	105.4	140.4	76.4	.000	-10.0
P23:50.00%	11.0	75.8	105.3	138.7	78.7	.000	-10.0
P23:J24	22.0	81.8	105.2	137.8	77.0	.000	-10.0
+ P24:J24	.0	81.8	105.2	137.8	77.0	.000	-10.0
P24:20.00%	16.4	84.5	105.1	140.1	80.9	.000	-10.0
P24:40.00%	32.8	87.2	104.9	140.2	82.4	.000	-10.0
P24:60.00%	49.2	89.8	104.8	138.2	79.8	.189	-10.0
P24:80.00%	65.6	92.5	104.7	137.8	82.5	.664E-01	-10.0
P24:J28	82.0	95.2	104.5	134.7	95.2	.000	-10.0
+ P25:J28	.0	95.2	104.5	134.7	95.2	18.4	-10.0
P25:33.33%	11.9	95.5	104.5	134.4	93.8	.000	-10.0
P25:66.67%	23.7	95.7	104.4	121.1	95.0	.000	-10.0
P25:N1	35.6	95.9	104.3	104.3	104.3	.000	-10.0
+ P0:J1	.0	40.6	40.6	40.6	40.6	.000	-10.0
P0:J2	.5	35.5	40.6	40.6	40.6	.000	-10.0
+ P1:J2	.0	35.5	40.6	40.6	40.6	.000	-10.0
P1:J3	.5	35.5	40.6	40.6	40.6	.000	-10.0

7.3.2 Comparison of Present Research Results with other Expert's Research

Comparison of present research results (water hammer software modeling and SPSS modeling), with other expert's research results, shows similarity and advantages:

7.3.2.1 Wylie, E. B., and Streeter, V. L., 1982

Classical water hammer theory neglects convective terms, and assumes fluid wave speed, c (m/s) depends on the support conditions of the pipes. Among these methods, MOC-based schemes are most popular because these schemes provide the desirable attributes of accuracy, numerical efficiency and programming simplicity. This method has been used in present research [12].

7.3.2.2 Arris S Tijsseling, Alan E Vardy, 2002

Present research assumed 3 states in Field Tests; Transmission Line with surge tank and Water hammer in leakage and no leakage condition.Comparison shows similarity in results[13].

7.3.2.3 Ghidaoui and, Leyn et al., 2005

The efficiency of a model is a critical factor for Real-time Control (RTC), since several simulations are required within a control loop in order to optimize the control strategy.

Small simulation time steps are needed to reproduce the rapidly varying hydraulics. Comparison shows similarity in with our results [14].

7.3.2.4   Arturo S. Leon, 2007

Comparison shows similarity between present research results and the results observed by Arturo S. Leon, 2007 [15].

7.3.2.5   Apoloniusz Kodura, Katarzyna Weinerowska, 2005

Detailed conclusions drawn on the basis of experiments and calculations for the pipeline with a local leak are similar to the results observed by Kodura and Weinerowska, 2005[16-17] (Figure 7.2).

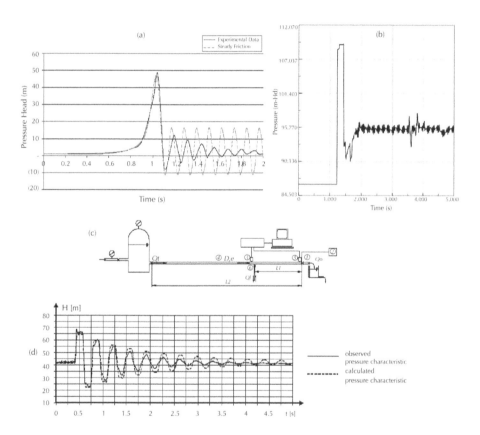

FIGURE 7.2   (a) Pressure head histories for a single pipe system, using steady and unsteady Friction. (Arturo S. Leon Research, 2007). (b) Rasht city water pipeline. (c) Pipeline with local leak, (d) Example of the measured and calculated pressure characteristics for the pipeline with local leak

7.4 CONCLUSION

In the computational technique presented, relationship between different variables accordance to fluid transient as a "Dynamic" operating is presented. Results are explained clearly in the section of "Model Summary and Parameter Estimates" as well as (Figure 7.3) and (Table 7.2–7.7).

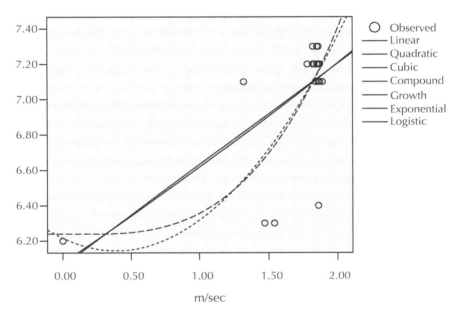

FIGURE 7.3 "Model Summary and Parameter Estimates" for Disinfection Water Networks and Transmission Lines.

KEYWORDS

- **Condition base maintenance**
- **Dynamic operating**
- **Fluid-structure interpenetration**
- **Method of characteristics**
- **Navier–Stokes equations**
- **Rasht city main pipeline**
- **Real-time control**
- **Water hammer models**

REFERENCES

1. Neshan, H. (1985). *Water Hammer*. pumps Iran Co. Tehran, Iran, pp. 1–60.

2. Streeter, V. L. and Wylie, E. B. (1979). *Fluid Mechanics*. McGraw-Hill Ltd., USA, pp. 492–505.

3. Hariri, K (2007a). Decreasing of unaccounted for water "UFW" by Geographic Information System "GIS" in Rasht Urban Water System. Technical and Art. *J. Civil Engineering Organization of Gilan* **38**, 3–7.

4. Hariri, K. (2007b). GIS and water hammer disaster at earthquake in Rasht water pipeline. 3rd International Conference on Integrated Natural Disaster Management (INDM) .Tehran, Iran.

5. Hariri, K. (2007c). Interpenetration of two fluids at parallel between plates and turbulent moving in pipe. 8th Conference on Ministry of Energetic works at research week. Tehran, Iran.

6. Hariri, K. (2007d). Water hammer and valves. 8th Conference on Ministry of Energetic works at research week. Tehran, Iran.

7. Hariri, K. (2007e). Water hammer and hydrodynamics' instability. 8th Conference on Ministry of Energetic works at research week. Tehran, Iran.

8. Hariri, K., (2007f). Water hammer analysis and formulation. 8th Conference on Ministry of Energetic works at research week. Tehran, Iran.

9. Hariri, K. (2007g). Water hammer and fluid condition. 8th Conference on Ministry of Energetic works at research week. Tehran, Iran.

10. Hariri, K. (2007h). Water hammer and pump pulsation. 8th Conference on Ministry of Energetic works at research week. Tehran, Iran.

11. Hariri, K. (2007i). Reynolds number and hydrodynamics' instability. 8th Conference on Ministry of Energetic works at research week. Tehran, Iran.

12. Wylie, E. B. and Streeter, V. L. (1982). Fluid Transients, Feb Press, Ann Arbor, MI, 1983. corrected copy: 166–171.

13. Arris, S. Tijsseling (in press). "Alan E Vardy Time scales and FSI in unsteady liquid-filled pipe flow"5–12.

14. Ghidaoui León et al. (2005). An efficient second-order accurate shock-capturing scheme for modeling one and two-phase water hammer flows. Ph. D. Thesis, 4–44.

15. Arturo, S. Leon (2007). An efficient second-order accurate shock-capturing scheme for modeling one and two-phase water hammer flows. Ph. D. Thesis, 4–44.

16. Apoloniusz, Kodura, Katarzyna, and Weinerowska (2005). Some Aspects of Physical and Numerical Modeling of Water Hammer in Pipelines, pp. 126–132.

17. Kodura, A., Weinerowska, K. (2005). Some aspects of physical and numerical modeling of water hammer in pipelines. In *International symposium on water management and hydraulic engineering*, pp. 125–33.

8 Improved Numerical Modeling for Perturbations in Homogeneous and Stratified Flows

CONTENTS

NOMENCLATURES

v_Φ = speed of long waves $\left(m/_s \right)$

λ = wavelength

σ = surface tension

h_0 = free surface of the liquid

h = liquid level is above the bottom of the channel

ξ = difference of free surface of the liquid and the liquid level is above the bottom of the channel(a deviation from the level of the liquid free surface)

u = fluid velocity $\left(m/_s \right)$

τ = time period

a = distance of the order of the amplitude

f = wavelength

k = wave number

Φ = a function of frequency and wave vector

$v_{\partial}(k)$ = phase velocity or the velocity of phase fluctuations $\left(m/_s \right)$

$\lambda(k)$ = wavelength

$\omega_{**}(k)$ = damping the oscillations in time

ω = waves with stationary in time but varying in length amplitudes

k = waves with a uniform length, but a time-varying amplitude.

8.1 INTRODUCTION

If on the surface of the deep pool filled with water to create a disturbance, then the surface of the water will begin to propagate. Their origin is explained by the fact that the fluid particles are located near the cavity. They create disturbances which will seek to fill the cavity under the influence of gravity. The development of this phenomenon is led to the spread of waves on the water. This chapter formulated the perturbations in homogeneous and stratified flows by improved numerical modeling.

One of the problems in the study of fluid flow in plumbing systems is the behavior of stratified fluid in the channels. Mostly steady flows initially are ideal, then the viscous and turbulent fluid in the pipes. The origin of propagation due to disturbance in deep pool which filled with water is explained by the fact that the fluid particles are located near the cavity. They create disturbances which will seek to fill the cavity under the influence of gravity. The development of this phenomenon is led to the spread of waves on the water. The fluid particles in such a wave do not move up and down around in circles. The waves of water are neither longitudinal nor transverse. They seem to be a mixture of both. The radius of the circles varies with depth of moving fluid particles. They reduce to as long as they do not become equal to zero. If we analyze the propagation velocity of waves on water, it will be reveal that the velocity of waves depends on length of waves.

8.2 MATERIALS AND METHODS

The speed of long waves is proportional to the square root of the acceleration of gravity multiplied by the wavelength $v_\Phi = \sqrt{g\lambda}$. The cause of these waves is the force of gravity. For short waves the restoring force due to surface tension force, and therefore the speed of these waves is proportional to the square root of the private. The numerator of which is the surface tension, and in the denominator—the product of the wavelength to the density of water $v_\Phi = \sqrt{\sigma/\lambda\rho}$.

Suppose there is a channel with a constant slope bottom, extending to infinity along the axis Ox. And let the feed in a field of gravity flows, incompressible fluid. It is assumed that the fluid is devoid of internal friction. Friction neglects on the sides and bottom of the channel. The liquid level is above the bottom of the channel h. A small quantity compared with the characteristic dimensions of the flow, the size of the bottom roughness, and so on. Let $h = \xi + h_0$, where h_0—ordinate denotes the free surface of the liquid (Figure 8.1). Free liquid surface h_0 (Figure 8.1), which is in equilibrium in the gravity field is flat. As a result of any external influence, liquid surface in a location removed from its equilibrium position. There is a movement spreading across the entire surface of the liquid in the form of waves, called gravity. They are caused by the action of gravity field. This type of waves occurs mainly on the liquid surface. They capture the inner layers, the deeper for the smaller liquid surface [1].

We assume that the fluid flow is characterized by a spatial variable x and time dependent t. Thus, it is believed that the fluid velocity u has a nonzero component u_x which will be denoted by u (other components can be neglected) in addition, the level of h depends only on x and t.

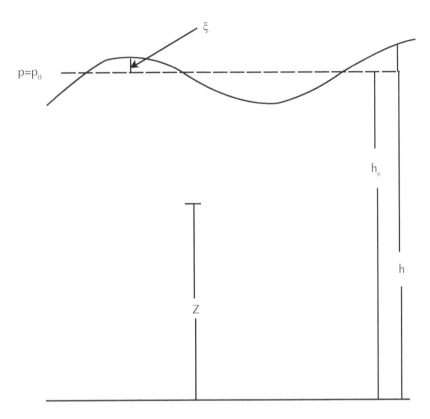

FIGURE 8.1 Schematic showing the layer of fluid of variable depth, where h_0 is the level of the free surface, ξ—a deviation from the level of the liquid free surface, h—depth of the fluid and z—vertical coordination of any point in the water column.

Let us consider such gravitational waves, in which the speed of moving particles are so small that for the Euler equation, it can be neglected the $(u\nabla)u$ compared with $\partial u/\partial t$. During the time period τ, committed by the fluid particles in the wave, these particles pass the distance of the order of the amplitude a. Therefore, the speed of their movement will be $u \sim a/\tau$. Rate u varies considerably over time intervals of the order τ and for distances of the order λ along the direction of wave propagation, λ—wavelength.

Therefore, the derivative of the velocity time—order u/τ and the coordinates—order u/λ. Thus, the condition $(u\nabla)u < \partial u/\partial t$ equivalent to the requirement

$$\frac{1}{\lambda}\left(\frac{a}{\lambda}\right)^2 << \frac{a}{\tau}\frac{1}{\tau} \quad a << \lambda \text{ or,} \tag{1}$$

that is amplitude of the wave must be small compared with the wavelength. Consider the propagation of waves in the channel Ox directed along the axis for fluid flow along the channel. Channel cross section can be of any shape and change along its length with changes in liquid level, cross-sectional area of the liquid in the channel denoted by $h = h(x,t)$. The depth of the channel and basin are assumed to be small compared with the wavelength. We write the Euler equation in the form of

$$\frac{\partial u}{\partial t} = -\frac{1}{\rho}\frac{\partial p}{\partial x} \tag{2}$$

$$\frac{1}{\rho}\frac{\partial p}{\partial z} = -g \tag{3}$$

where ρ —density, p —pressure g —acceleration of free fall. Quadratic in velocity members omitted, since the amplitude of the waves is still considered low [2].

From the second equation we have that at the free surface $z = h(x,t)$ (where) $p = p_0$ should be satisfied:

$$p = p_0 + \rho g(h - z) \tag{4}$$

Substituting this expression in equation (2), we obtain to determine u and h we use the continuity equation for the case under consideration.

$$\frac{\partial u}{\partial t} = -g\frac{\partial h}{\partial x} \tag{5}$$

Consider the volume of fluid contained between two planes of the cross-section of the canal at a distance dx from each other. Per unit time through a cross-section x enter the amount of fluid, equal to $(hu)_x$. At the same time through the section $x + dx$ there is forth coming $(hu)_{x+dx}$. Therefore, the volume of fluid between the planes is changed to

$$(hu)_{x+dx} - (hu)_x = \frac{\partial(hu)}{\partial x}dx$$

By virtue of incompressibility of the liquid is a change could occur only due to changes in its level. Changing the volume of fluid between these planes in a unit time is equal

$$\frac{\partial h}{\partial t}dx$$

Consequently, we can write:

$$\frac{\partial(hu)}{\partial x}dx = -\frac{\partial h}{\partial t}dx \text{ and } \frac{\partial(hu)}{\partial x} + \frac{\partial h}{\partial t} = 0, t > 0, -\infty < x < \infty \text{ or,} \tag{6}$$

Since $h = h_0 + \xi$ where a h_0 —denotes the ordinate of the free liquid surface (Figure 8.1), in a state of relative equilibrium and evolving the influence of gravity is

$$\frac{\partial \xi}{\partial t} + h_0\frac{\partial u}{\partial x} = 0 \tag{7}$$

Thus, we obtain the following system of equations describing the fluid flow in the channel:

$$\frac{\partial \xi}{\partial t} + h_0 \frac{\partial u}{\partial x} = 0, \frac{\partial u}{\partial t} + g \frac{\partial \xi}{\partial x} = 0, t > 0, -\infty < x < \infty \qquad (8)$$

8.3 RESULTS AND DISCUSSION

8.3.1 Velocity Phase of the Harmonic Wave

The phase velocity h_0 expressed in terms of frequency v_Φ and wavelength f (or the angular frequency) λ and wave number $\omega = 2\pi f$ formula $k = 2\pi / \lambda$. The concept of phase velocity can be used if the harmonic wave propagates without changing shape. This condition is always performed in linear environments. When the phase velocity depends on the frequency, it is equivalent to talk about the velocity dispersion. In the absence of any dispersion the waves assumed with a rate equal to the phase velocity. Experimentally, the phase velocity at a given frequency can be obtained by determining the wavelength of the interference experiments. The ratio of phase velocities in the two media can be found on the refraction of a plane wave at the plane boundary of these environments. This is because the refractive index is the ratio of phase velocities. It is known that the wave number k satisfies the wave equation are not any values ω but only if their relationship. To establish this connection is sufficient to substitute the solution of the form $\exp\left[i(\omega t - kx)\right]$ in the wave equation [3]. The complex form is the most convenient and compact. We can show that any other representation of harmonic solutions, including in the form of a standing wave leads to the same connection between ω and k. Substituting the wave solution into the equation for a string, we can see that the equation becomes an identity for $\omega^2 = k^2 v_\Phi^2$. Exactly the same relation follows from the equations for waves in the gas, the equations for elastic waves in solids and the equation for electromagnetic waves in vacuum.

8.4 CONCLUSION

The presence of energy dissipation, [4] leads to the appearance of the first derivatives (forces of friction) in the wave equation. The relationship between frequency and wave number becomes the domain of complex numbers. For example, the telegraph equation (for electric waves in a conductive line) yields $\omega^2 = k^2 v_\Phi^2 + i \cdot \omega R / L$. The relation connecting between a frequency and wave number (wave vector), in which the wave equation has a wave solution is called a dispersion relation, the dispersion equation or dispersion. This type of dispersion relation determines the nature of the wave. Since the wave equations are partial differential equations of second order in time and coordinates, the dispersion is usually a quadratic equation in the frequency or wave number. The simplest dispersion equations presented above for the canonical wave equation are also two very simple solutions $\omega = +k v_\Phi$ and $\omega = -k v_\Phi$. We know that these two solutions represent two waves traveling in opposite directions. By its physical meaning the frequency is a positive value so that the two solutions must define two values of the wave number, which differ in sign. The Act permits the dispersion, generally speaking, the existence of waves with all wave numbers that is of any length, and, consequently, any frequencies. The phase velocity of these waves $v_\Phi = \omega/k$ coincides with the most velocity, which appears in the wave equation and is a constant which depends only on the properties of the medium. The phase velocity depends on the wave number, and, consequently, on the frequency. The dispersion equation for the telegraph equation is an algebraic quadratic equation has complex roots. By analogy with the theory of oscillations, the presence of imaginary part of the frequency means the damping or growth of waves. It can be noted that the form of the dispersion law determines the presence of damping or growth. In general terms, the dispersion can be represented by the equation $\Phi(\omega, \vec{k}) = 0$ where Φ—a function of frequency and wave vector. By solving this equation for ω you can obtain an expression for the phase velocity) $v_\Phi = \omega/k = f(\omega, \vec{k})$. By definition, the phase velocity is a vector directed normal to phase surface. Then, more correctly write the last expression in the following form:

$$\vec{v_\Phi} = \frac{\lambda}{T} = \frac{\omega}{k^2} \cdot \vec{k} = f(\omega, \vec{k})$$

Dispersive properties of media the most important subject of research in wave physics, which has the primary practical significance. If we refer to dimensionless parameters and variables:

$$\tau = t\sqrt{\frac{g}{h_0}}, \; X = \frac{x}{h_0}, \; U = u\frac{1}{\sqrt{gh_0}}, \; \delta = \frac{\xi}{h_0},$$

the system of equations (2.4.8) becomes

$$\frac{\partial \delta}{\partial \tau} + \frac{\partial U}{\partial X} = 0, \; \frac{\partial U}{\partial \tau} + \frac{\partial \delta}{\partial X} = 0, \; t > 0, \; -\infty < X < \infty \tag{9}$$

Consider a plane harmonic longitudinal waves that is, we seek the solution of (9) as the real part of the following complex expressions:

$$\Psi = \Psi^0 \exp\left[i(k_* X + \omega_* \tau)\right], \; \Psi^0 = \Psi_*^0 + i\Psi_{**}^0, \; |\Psi^0| \ll 1$$

$$k_* = k + ik_{**}, \; \omega_* = \omega + i\omega_{**} \tag{10}$$

where $\Psi = \delta, U$, a $\Psi^0 = \delta^0, U^0$ determines the amplitude of the perturbations of displacement and velocity.

There are two types of solutions.

Type I. Solution, or wave of the first type, when $k_* = k$—a real positive number $(k > 0, k_{**} = 0)$.

In this case we have:

$$\Psi = \left(\Psi_*^0 + i\Psi_{**}^0\right)\exp\left[i\left(kX + \omega\tau + i\omega_{**}\tau\right)\right] = \left(\Psi_*^0 + i\Psi_{**}^0\right)\exp\left(-\omega_{**}\tau\right)\times$$

$$\left[\cos\left(kX + \omega\tau\right) + i\sin\left(kX + \omega\tau\right)\right]$$

$$\mathrm{Re}\{\Psi\} = \exp\left(-\omega_{**}\tau\right)\left|\Psi^0\right|\sin\left[\varphi + \left(kX + \omega\tau\right)\right]$$

$$\left|\Psi^0\right| = \sqrt{\Psi_*^{0\,2} + \Psi_{**}^{0\,2}}, \quad \varphi = arctg\left(-\Psi_*^0 / \Psi_{**}^0\right)$$

Here

Thus, the decision of the first type is a sinusoidal coordinate and $\omega_{**} > 0$ decaying exponentially in time perturbation, which is called k —wave:

$$\Psi(k) = \left|\Psi^0\right|\exp\left[-\omega_{**}(k)\tau\right]\sin\left\{\varphi + \frac{2\pi\left[X + v_\Phi(k)\tau\right]}{\lambda(k)}\right\} \tag{11}$$

Where $v_\Phi(k) = \omega(k)/k$, $\lambda(k) = 2\pi/k$,) φ—initial phase.

Here, $v_\Phi(k)$—phase velocity or the velocity of phase fluctuations $\lambda(k)$—wavelength) $\omega_{**}(k)$—damping the oscillations in time. In other words, k —waves - waves have uniform length, but time-varying amplitude. These waves are analogue of free oscillations.

Type II. Decisions, or wave, the second type, when $\omega_* = \omega$—a real positive number $(\omega > 0, \omega_{**} = 0)$. In this case we have:

$$\Psi = \left(\Psi_*^0 + i\Psi_{**}^0\right)\exp\left[i\left(kX + \omega\tau + ik_{**}z\right)\right] = \left(\Psi_*^0 + i\Psi_{**}^0\right)\exp\left(-k_{**}X\right)\times$$

$$\left[\cos\left(kX + \omega\tau\right) + i\sin\left(kX + \omega\tau\right)\right]$$

$$\mathrm{Re}\{\Psi\} == \exp\left(-k_{**}X\right)\left|\Psi^0\right|\sin\left[\varphi + \left(kX + \omega\tau\right)\right]$$

Thus, the solution of the second type is a sinusoidal oscillation in time (excited, for example, any stationary source of external monochromatic vibrations at) $X = 0$, decaying exponentially along the length of the amplitude. Such disturbances, which are analogous to a wave of forced oscillations, called ω—waves:

$$\Psi(\omega) = \left|\Psi^0(\omega)\right|\exp\left(-k_{**}(\omega)X\right)\sin\left\{\varphi + \frac{2\pi\left[X + v_\Phi(\omega)\tau\right]}{\lambda(\omega)}\right\} \tag{12}$$

$$v_\Phi(\omega) = \omega/k(\omega), \lambda(\omega) = 2\pi/k(\omega)$$

Here, $k_{**}(\omega)$—damping vibrations in length. In other words, ω—waves - waves with stationary in time but varying in length amplitudes. Cases $k < 0, k_{**} > 0$ and $k > 0, k_{**} < 0$ consistent with attenuation of amplitude of the disturbance regime in the direction of phase fluctuations or phase velocity.

Let us obtain the characteristic equation, linking k_* and ω_*. After substituting (10) in the system of equations (9) we obtain:

$$\delta^0 \frac{\omega_*}{k_*} + U^0 = 0$$

$$U^0 \frac{\omega_*}{k_*} + \delta^0 = 0 \tag{13}$$

From the condition of the existence of a system of linear homogeneous algebraic equations (13) with respect to perturbations of a nontrivial solution implies the desired characteristic, or dispersion, which has one solution:

$$v_\Phi = \sqrt{gh_0} \tag{14}$$

Thus, we obtain a solution representing a sinusoidal in time and coordinate free undammed oscillations. Such behaviors of the waves are due to the absence of any dissipation in the fluid. The fluid is incompressible and ideal. There is no heat–mass transfer. Equations (9) with respect to perturbations take the form of wave equations:

$$\frac{\partial^2 \xi}{\partial t^2} = gh_0 \frac{\partial^2 \xi}{\partial x^2} \quad \text{and} \quad \frac{\partial^2 u}{\partial t^2} = gh_0 \frac{\partial^2 u}{\partial x^2} \tag{15}$$

Note that in gas dynamics $v_\Phi = \sqrt{gh_0}$ equivalent to the speed of sound.

KEYWORDS

- **Euler equation**
- **Gravity field**
- **Plumbing systems**
- **Quadratic equation**
- **Telegraph equation**
- **Wavelength**

REFERENCES

1. Nagiyev, F. B. (1993). *Dynamics, heat and mass transfer of vapor-gas bubbles in a two-component liquid.* Turkey-Azerbaijan petrol semin, Ankara, Turkey.
2. Nagiyev, F. B. (1995). The method of creation effective coolness liquids. Third Baku international Congress. Baku, Azerbaijan Republic, September 19–22.

3. Nigmatulin, R. I., Khabeev, N. S., and Nagiyev, F. B. (1981). Dynamics, heat and mass transfer of vapor-gas bubbles in a liquid. *Int. J. Heat Mass Transfer* **24**(6), Printed in Great Britain, pp. 1033–1044.

4. Loytsyanskiy, L. G. (1970). Fluid, Moscow, Nauka, p. 904.

9 Computational Model for Water Hammer Disaster

CONTENS

NOMENCLATURES

t = time (s)

λ_0 = unit of length

p = pressure (N/m2)

V = velocity (m/s)

S = length (m)

D = diameter of each pipe (m)

R = pipe radius (m)

γ = specific weight (N/m3)

v = fluid dynamic viscosity (kg/m.s)

hp = head gain from a pump (m)

hL = combined head loss (m)

P = surge pressure (m)

ρ = density (kg/m3)

C = velocity of surge wave (m/s)

P/γ = pressure head (m)

Z = elevation head (m)

$V 2/2$ = velocity head (m)

P = pressure (N/m2),

γ = specific weight, (N/m3)

Z = elevation (m)

Hp = surge wave head at pipeline points- intersection points of characteristic lines (m)

Vp = surge wave velocity at pipeline points- intersection points of characteristic lines (m/s)

G = gravitational acceleration constant (m/s2)

Vri = surge wave velocity at right hand side of intersection points of characteristic lines (m/s)

Hri = surge wave head at right hand side of intersection points of characteristic lines (m)

Vle = surge wave velocity at left hand side of intersection points of characteristic lines (m/s)

Hle = surge wave head at left hand side of intersection points of characteristic lines (m)

C⁻ = characteristic lines with negative slope

C+ = characteristic lines with positive slope

Min = Minimum

Max = Maximum

Lab = Laboratory

9.1   INTRODUCTION

This chapter presents the application of computational performance of a numerical method by a dynamic model. The model has been presented by method of the Eulerian based expressed in a method of characteristics (MOC). It has been defined by finite difference form for heterogeneous model with varying state in the system. Present work offered MOC as a computational approach from theory to practice in numerical analysis modeling. Therefore, it is computationally efficient for transient flow irreversibility prediction in a practical case. In this work reclamation numerical analysis modeling showed the lining method as the best construction way for reclamation of damaged water transmission line.

Irreversibility as a fluid dynamics phenomenon is an important case study about heterogeneous model with varying state within the system for designer engineers. Water hammer as an effect of irreversibility of fluid is a pressure surge or wave. It is generated by the kinetic energy of a fluid in motion when it is forced to stop or change direction suddenly [1]. The majority of pressure transients in water and waste water systems are the result of changes at system boundaries. It is revealed typically at the upstream and downstream ends of the system or at   highest points of water transmission line. Consequently, Results of present work reduce the risk of system damage or failure. With proper analysis it determines the system's default dynamic response. Design of protection equipment helps to control transient energy. It specifies operational procedures to avoid transients [2]. Various methods have been developed to solve transient flow in pipes. These ranges are included from approximate equations to numerical solutions of the nonlinear Navier–Stokes equations. Hydraulic transient flow is also known as unsteady fluid flow. During a transient analysis, the fluid and system boundaries can be either elastic or inelastic [3]. This work investigated reclamation numerical analysis about unaccounted for water "UFW" at water pipeline. So a suitable model was introduced for this proposes. The location and rate of water release from pipeline was predicted by the leakage location predictor modeling. Thus defined model was used for selection of the optimized way at make decision on reclamation numerical analysis for damaged water transmission line. In this work, according to pipeline specification by applying MOC three models were defined. Thus the leakage point as the most important and as a critical point was selected for analysis by three models of transient flow [4].

9.2 MATERIALS AND METHODS

9.2.1 Experimental (Field Tests Model Criteria)

In this work water hammer numerical modeling and simulation processes was handled by nonlinear heterogeneous model.

9.2.2 Approaches to Transient Flow (Method of characteristics "MOC" Model)

Model was defined by MOC (Table 9.1). In this work water hammer software was applied for numerical modeling (Table 9.2) .Specification and geography of system were defined for water hammer software, Version 07.00.049.00. In this case, water pipeline assumed in water leakage condition [5] and equipped with surge tank (real condition or existent condition).

The MOC is based on a finite difference technique where pressures are computed along the pipe for each time step (Joukowski, 1904). The combined elasticity of both the water and the pipe walls is characterized by the pressure wave speed (Arithmetic method [3] combination of the Joukowski formula and Allievi formula, Wylie & Streeter, 1982).

The MOC approach transforms the water hammer partial differential equations into the ordinary differential equations along the characteristic lines (Figure 8.1).

Balancing the energy across two points in the system yields the energy or Bernoulli equation for steady-state flow. The components of the energy equation can be combined to express two useful quantities, the hydraulic grade and the energy grade:

$$\left(P_1/\gamma\right)+Z_1\left(V_1^2/2g\right)+h_p=\left(P_2/\gamma\right)+Z_2+\left(V_2^2/2g\right)+h_L, \tag{1}$$

$$(g/a)(dH/dt)+dv/dt+\left(f\,v|v|2d\right)=0\Rightarrow(ds/dt)=c^+, \tag{2}$$

$$-(g/a)(dH/dt)+dv/dt+\left(f\,v|v|2d\right)=0\Rightarrow(ds/dt)=c^-, \tag{3}$$

The method of characteristics is a finite difference technique where pressures were computed along the pipe for each time step. Calculation automatically sub-divided the pipe into sections (intervals) and selected a time interval for computations.

$$(dp/dt) = (\partial p/\partial t) + (\partial p/\partial s)(ds/dt), \tag{4}$$

$$(dv/dt) = (\partial v/\partial t) + (\partial v/\partial s)(ds/dt), \tag{5}$$

P and V changes due to time are high and due to coordination are low then it can be neglected for coordination differentiation:

$$(\partial v/\partial t) + (1/\rho)(\partial p/\partial s) + g(dz/ds) + (f/2D)v|v| = 0, \text{ (Euler equation)} \tag{6}$$

$$C^2(\partial v/\partial s) + (1/P)(\partial P/\partial t) = 0, \text{ (Continuity equation),} \tag{7}$$

By linear combination of Euler and continuity equations in characteristic solution Method:

$$\lambda\left[(\partial v/\partial t) + (1/\rho)(\partial p/\partial s) + g(dz/ds) + (f/2D)v|v|\right] + C^2(\partial v/\partial s) + (1/p)(\partial p/\partial t) = 0, \tag{8}$$
$$\lambda =^+ c \,\&\, \lambda =^- c$$

$$(dv/dt) + (1/cp)(dp/ds) + g(dz/ds) + (f/2D)v|v| = 0, \tag{9}$$

$$(dv/dt) - (1/cp)(\partial p/\partial s) + g(dz/ds) + (f/2D)v|v| = 0, \tag{10}$$

Method of characteristics drawing in (s-t) coordination:

$$(dv/dt) - (g/c)(dH/dt) = 0, \tag{11}$$

$$dH = (c/g)dv, \text{ (Joukowski Formula),} \tag{12}$$

By Finite Difference method:

$$c+ : \left((vp - v_{Le})(Tp - 0)\right) + \left((g/c)(Hp - H_{Le})/(Tp - 0)\right) + \left((\ fv_{Le}|\ v_{L\ e}|\ )/2D\right) = 0| \tag{13}$$

$$c- : \left((vp - vRi)(Tp - 0)\right) + \left((g/c)(Hp - HRi)/(Tp - 0)\right) + \left((\ fvRi|vRi|\ )/2D\right) = 0| \tag{14}$$

$$c+ : (vp - v_{Le}) + (g/c)(Hp - H_{Le}) + (f\Delta t)\left(fv_{Le}|v_{Le}|\right)/2D) = 0 \tag{15}$$

$$c- : (vp - vRi) + (g/c)(Hp - HRi) + (f\Delta t)\left(v_{Ri}|v_{Ri}|\right)/2D) = 0 \tag{16}$$

$$V_p = 1/2 \begin{pmatrix} (V_{Le} + V_{ri}) + (g/c)(H_{Le} - H_{ri}) \\ -(f\ \Delta t/2D)(V_{Le}\ |V_{Le}| + V_{ri}|V_{ri}|) \end{pmatrix}, \tag{17}$$

$$H_P = 1/2 \begin{pmatrix} C/g(V_{Le} - V_{ri}) + (H_{Le} + H_{ri}) \\ -C/g(f \, \Delta t / 2D)(V_{Le} \, |V_{Le}| - V_{ri} |V_{ri}|) \end{pmatrix}, \tag{18}$$

9.3 RESULTS AND DISCUSSION

Local leakage effect on high pressure drop in water pipeline. For make decision about reclamation of leakage in water transmission line, it was investigated the water hammer modeling for three cases at leakage point [6-10]:

- Simulation of pipeline (Figure 8.2) for reclamation numerical analysis of water transmission line by replace of present reinforced concrete pipe (AC pipe) with the same diameter of polyethylene pipe (AC pipe-1,200 mm replace with PE pipe-1,200 mm).
- Simulation of pipeline (Figure 8.3) for reclamation numerical analysis of water transmission line by lining (push a new smaller diameter pipe into the old larger diameter pipe- Push PE pipe-1,100 mm into AC pipe-1,200 mm).
- Simulation of pipeline (Figure 8.3) for reclamation numerical analysis of water transmission line by replace of present reinforced concrete pipe (AC pipe) with the larger diameter of polyethylene pipe (AC pipe-1,200 mm replace with PE pipe-1,300 mm).

TABLE 9.1   Heterogeneous Model (Software Hammer Version 07.00.049.00).

Type	Keyword	Value
1	Title	numerical Modeling of fluid interaction by Method of characteristics "MOC"-
		Rasht city water pipeline in the north of Iran.
2	Units for Flow	cms
3	Units for Head	m
4	Units for Volume	m3
5	Units for Diameter	mm
6	Units for Length	m
7	Units for Mass	kg
8	Time Increment	0.0148
9	Number of Time Steps	339
10	Simulation Time	5.003
11	Wave Speed	1084
12	Vapour Pressure	-10
13	File Name	E:\k-hariri Asli\ surge tank-leakage-inp
14	Date of Run	10/10/2008
15	Time of Run	11:09.4
16	Number of Nodes	27

TABLE 9.1 *(Continued)*

Type	Keyword	Value
17	Number of Pipes	26
18	Licensee Name	HMI
19	Licensee Address	Waterbury, CT
20	EHG Name	Haestad Methods, Inc.
21	EHG Address	Waterbury, CT
22	Units for Force	N
23	Force Reports	No
24	Volume Scale Factor	1
25	Flow Scale Factor	1
26	Labels	Short

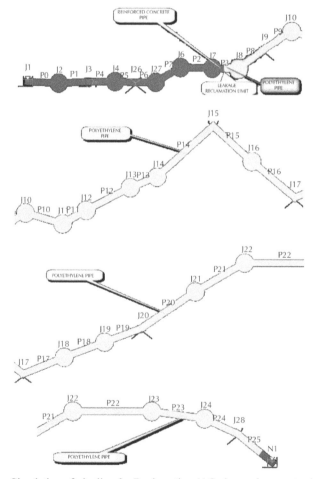

FIGURE 9.1 Simulation of pipeline for Reclamation (AC pipe replacement with PE pipe).

TABLE 9.2   Heterogeneous model pipes.

Pipe	Length m)	Diameter (mm)	Velocity (m/s)	Hazen-Williams Friction Coef.
P3	311	1200	2.21	90
P4	1	1200	2.65	90
P5	0.5	1200	2.65	91
P6	108.7	1200	2.65	67
P7	21.5	1200	2.21	90
P8	15	1200	2.21	86
P9	340.7	1200	2.21	90
P10	207	1200	2.21	90
P11	339	1200	2.21	90
P12	328.6	1200	2.21	90
P13	47	1200	2.21	90
P14	590	1200	2.21	90
P15	49	1200	2.21	90
P16	224	1200	2.21	90
P17	18.4	1200	2.21	90
P18	14.6	1200	2.21	90
P19	12	1200	2.21	90
P20	499	1200	2.21	90
P21	243.4	1200	2.21	90
P22	156	1200	2.21	90
P23	22	1200	2.21	90
P24	82	1200	2.21	90
P25	35.6	1200	2.21	90
P0	0.5	1200	2.65	90
P1	0.5	1200	2.65	90

TABLE 9.3   Field tests model (leakage location at extreme heads)

Point	Distance	Elev.	Init Head	Max Head	Min Head	Vapor Pr.	
+ P3:J7	.0	36.3	132.8	156.8	105.1	.000	-10.0
P3:5.00%	15.6	36.3	132.7	156.6	105.3	.000	-10.0
P3:10.00%	31.1	36.3	132.6	156.3	105.1	.000	-10.0
P3:15.00%	46.7	36.3	132.4	156.1	105.0	.000	-10.0
P3:20.00%	62.2	36.3	132.3	155.8	104.9	.000	-10.0
P3:25.00%	77.8	36.3	132.2	155.6	104.7	.000	-10.0
P3:30.00%	93.3	36.2	132.1	155.5	104.4	.000	-10.0
P3:35.00%	108.9	36.2	131.9	155.3	104.3	.000	-10.0
P3:40.00%	124.4	36.2	131.8	155.2	104.2	.000	-10.0
P3:45.00%	140.0	36.2	131.7	155.0	104.0	.000	-10.0
P3:50.00%	155.5	36.2	131.6	135.1	103.8	.000	-10.0
P3:55.00%	171.1	36.2	131.4	134.9	103.7	.000	-10.0

TABLE 9.3   *(Continued)*

Point	Distance	Elev.	Init Head	Max Head	Min Head	Vapor Pr.	
P3:60.00%	186.6	36.2	131.3	134.7	103.5	.000	-10.0
P3:65.00%	202.2	36.2	131.2	134.4	103.4	.000	-10.0
P3:70.00%	217.7	36.2	131.1	134.6	102.9	.000	-10.0
P3:75.00%	233.3	36.2	130.9	134.4	102.7	.000	-10.0
P3:80.00%	248.8	36.1	130.8	134.0	102.7	.000	-10.0
P3:85.00%	264.4	36.1	130.7	133.9	102.4	.000	-10.0
P3:90.00%	279.9	36.1	130.6	133.6	102.3	.000	-10.0
P3:95.00%	295.5	36.1	130.4	133.5	102.1	.000	-10.0

+ P3:J8 According to pipeline specification three models were defined. The leakage point (+ P3:J7) as the most important critical point (Table 9.3) was selected for analysis by above three models for pipeline reclamation numerical analysis. Comparison of these three cases revealed the reclamation numerical analysis curves as long as water transmission line at transient flow condition (Figure 9.2–9.5).

All of these numerical analysis curves confirmed the critical effect of leakage point for make decision about reclamation of present water transmission line. For all three models were showed the numerical analysis of existent pipe reclamation (reinforced concrete pipe replacement with the offered polyethylene pipe) as flowing procedure [11-12]:

1. *Max.* Pressure–Distance; *Min.* Pressure-Distance curves shows pressure decreased after the leakage point (+ P3:J7) .Pressure drop is high proportional to diameter increasing (Figure 9.2–9.5).
2. *Max.* Pressure–Distance curve shows pressure rising before the leakage point (+ P3:J7). Pressure rising became low and low proportional to diameter increasing (Figure 9.2–9.5).
3. *Min.* Pressure–Distance curve shows pressure decreasing before the leakage point (+ P3:J7). So pressure drop became low and low proportional to diameter increasing (Figure 9.2–9.5).
4. With polyethylene pipe diameter increasing at the leakage point (+ P3:J7), *Min.* Pressure and *Max.* Pressure drop happened at the near of pump station location (Figure 9.2–9.5).
5. *Max.* Pressure drop happened when diameter changed from small diameter to large diameter. *Min.* Pressure drop happened when diameter changed from large diameter to small diameter.
6. 5-pressure variation (Figure 9.2–9.5) interval decreased by diameter increasing at the leakage point (+ P3:J7).

Reclamation numerical analysis modeling showed the best construction way for water transmission line was the lining of present reinforced concrete pipe (AC pipe). This variant for reclamation was based on lining of present reinforced concrete pipe with the smaller diameter of polyethylene pipe (AC pipe-1,200 mm must be replaced by PE pipe-1,100 mm). It was the best construction way for reclamation. But the reclamation numerical modeling showed pressure drop happened when diameter changed from large diameter to smaller diameter. Many factors including: Total budgets—Time of project and and so on. were redound to selection of the two variants:

(a) Reclamation of water transmission line by lining, (b) Reclamation of water transmission line by replacement of existent pipe with the larger diameter of polyethylene pipe (AC pipe-1,200 mm replace with PE pipe-1,300 mm).

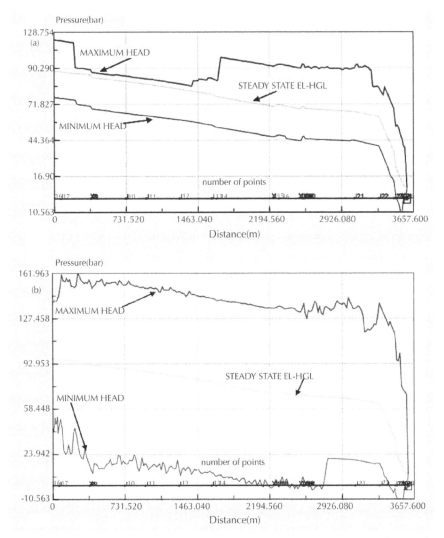

FIGURE 9.2   Simulation of pipeline for Reclamation: (a) AC pipe-1,200 mm, (b) AC pipe-1,200 mm replacement with PE pipe-1,200 mm.

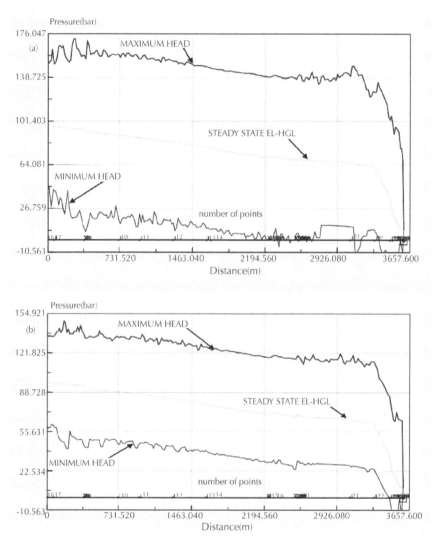

FIGURE 9.3 Simulation of pipeline for Reclamation: (a) AC pipe-1,200 mm replacement with PE pipe-1,100 mm, (b) AC pipe-1,200 mm replacement with PE pipe-1,300 mm.

9.4 CONCLUSION

Transient analysis should be performed for large, high-value pipelines, especially with pump stations. A complete transient analysis, in conjunction with other system design activities, should be performed during the initial design phases of a project. Normal flow-control operations and predicable emergency operations should, was evaluated during the work. Make decision process from theory to practice for reclamation of damaged water transmission line was considered in this work. It confirmed by nonlinear heterogeneous model for water hammer in three cases. Reclamation of unaccounted for water "UFW" as a water hammer effect was investigated for existent water transmission line.

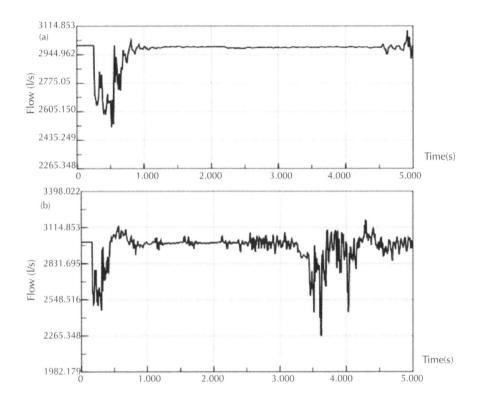

FIGURE 9.4 Simulation of pipeline for Reclamation: (a) AC pipe-1,200 mm, (b) AC pipe-1,200 mm replacement with PE pipe-1,200 mm.

Reclamation numerical modeling showed pressure decreasing became high and high proportional to diameter increasing. It also showed pressure decreasing became low and low proportional to diameter increasing. *Max.* Pressure drop happened when diameter changed from smaller diameter to larger diameter and *Min.* Pressure decreasing happened when diameter changed from larger diameter to smaller diameter. Pressure rising became low and low proportional to diameter increasing. *Max* pressure drop happened as long as diameter changed from small diameter to larger diameter and *Min* pressure drop happened when diameter changed from large diameter to smaller diameter. This was showed the numerical analysis modeling as a computational approach is computationally efficient for transient flow irreversibility prediction in a practical case. It offered the lining method as a construction way for reclamation of damaged water transmission line. The reason for offering of this variant for reclamation was based on the lining of present reinforced concrete pipe with smaller diameter of polyethylene pipe (AC pipe-1,200 mm must be replaced by PE pipe-1,100 mm).

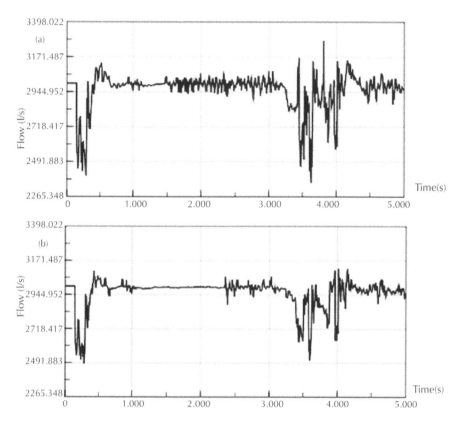

FIGURE 9.5 Simulation of pipeline for Reclamation: (a) AC pipe-1,200 mm replacement with PE pipe-1,100 mm, (b) AC pipe-1,200 mm replacement with PE pipe-1,300 mm.

9.4.1 Comparison of Present research results with other expert's research

[12], In the work of Kodura and Weinerowska [13] water hammer in pipeline at the local leak case have been presented. This case was related to some additional factors. Therefore detailed conclusions drawn on the basis of experiments and calculations for the pipeline with a local leak were presented in the paper of [13]. The most important points which were observed are as flowing:

The effects of discharge from local leak to total discharge in the pipeline were investigated. These effects were studied related to the values for period of oscillations. In a consequence it was studied related to the value of wave celerity when the outflow to the overpressure reservoir from the leak was imposed [13].

Important points which were mentioned by present work are as the flowing items:

9.4.2 Air Entrance Approaches

Analysis and comparison of nonlinear heterogeneous model for water hammer results showed that at point P24:J28 of water pipeline, air was interred to system. *Max.* volume of penetrated air was equal to 198.483(m³) and currently flow was equal to 2.666 (m³/s).

Local leakage rate effects on the total transmission flow. The influence of the ratio of discharge from local leak to the total discharge in the pipeline was affected by the values of oscillations period. In a consequence it was affected to the value of wave celerity when the outflow to the overpressure reservoir from the leak. Thus Present work was conformed to the results of the Kodura and Weinerowska's work (Figure 9.6).

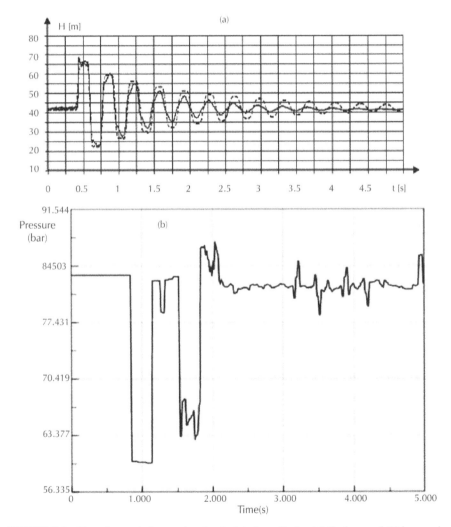

FIGURE 9.6 Experimental observed and calculated results for (a) Kodura and Weinerowska research, (b) present research.

KEYWORDS

- **Damaged water transmission line**
- **Heterogeneous model**
- **Numerical method**
- **Reclamation**
- **Simulation of pipeline**
- **Waste water systems**
- **Water hammer**

REFERENCES

1. Leon, S. A. (2007). Improved Modeling of Unsteady Free Surface. Pressurized and Mixed Flows in Storm-Sewer Systems, Submitted in Partial Fulfillment of the Requirements for the degree of Doctor of Philosophy in Civil Engineering in the Graduate College of the University of Illinois at Urbana-Champaign, 57–58.

2. Hariri Asli, K., Nagiyev, F. B., and Haghi, A. K. (2009a). Three-dimensional Conjugate Heat Transfer in Porous Media, International J. of the Balkan Tribological Association, Thomson Reuters Master Journal List, ISSN: 1310–4772. Sofia, Bulgaria, 336.

3. Hariri Asli, K., Nagiyev, F. B., and Haghi, A. K. (2009b). Computational Methods in Applied Science and Engineering Interpenetration of Two Fluids at Parallel between Plates and Turbulent Moving in Pipe, Nova Science, New York, USA, 115–128. https://www.novapublishers.com/catalog/ product_ info.php? products_id=10681.34

4. Wylie, E. B. and Streeter, V. L. (1982). Fluid Transients in Systems. Prentice Hall, 1993, corrected copy: 166–171.

5. Kodura, A. and Weinerowska, K. (2005). Some Aspects of Physical and Numerical Modeling of Water Hammer in Pipelines, pp. 125–133.

6. Hariri Asli, K., Nagiyev, F. B., and Haghi, A. K. (2009c). Some Aspects of Physical and Numerical Modeling of Water Hammer in Pipelines, Nonlinear Dynamics. *International J. of Nonlinear Dynamics and Chaos in Engineering Systems*, ISSN: 0924-090X (print version), ISSN: 1573-269X (electronic version), Journal No 11071 Springer, Published online: 10 December, http://nody.edmgr.com/

7. Hariri Asli, K., Nagiyev, F. B., and Haghi, A. K. (2009d). Computational Methods in Applied Science and Engineering. Water hammer analysis: Some computational aspects and practical hints Nova Science Publications, New York, USA.

8. Wood Don, J. (2005). Water hammer Analysis, Essential and Easy (and Efficient) *J. Envir.Engrg.* **131**, 1123, Canada.

9. Ming, Z. and Ghidaoui, S. M. (2004). Godunov-Type Solutions for Water Hammer Flows. *J. Hydr. Engrg.* **130**, 341.

10. Stephenson (2004). Closure to Simple Guide for Design of Air Vessels for Water Hammer protection of pumping Lines. *J. Hydraulic Engineering* **130**, 275.

11. Brunone, B., Bryan, W., Karney, M., and Mecarelli, M. (2000). Ferrante: Velocity profiles and Unsteady Pipe Friction in Transient Flow. *J. Water Resour. Plng. and Mgmt.* **126**, 236, Lublin.

12. Apoloniusz, Kodura, Katarzyna, and Weinerowska (2005). Some Aspects of Physical and Numerical Modeling of Water Hammer in Pipelines, International symposium on water manage-

ment and hydraulic engineering, 4–7th September, paper No.11.05, pp. 126–132, Ottenstein, Austria.

13. Kodura, A. and Weinerowska, K. (2005). Some aspects of physical and numerical modeling of water hammer in pipelines, International symposium on water management and hydraulic engineering, 4–7th September, paper No.11.05, pp. 125–133, Ottenstein, Austria.

10 Heat and Mass Transfer in Binary Mixtures; A Computational Approach

CONTENTS

NOMENCLATURES

R_0 = radiuses of a bubble

D = diffusion factor

β = cardinal influence of componential structure of a mixture

N_{k_0}, N_{c_0} = mole concentration of 1-th component in a liquid and steam

γ = Adiabatic curve indicator

c_l, c_{pv} = specific thermal capacities of a liquid at constant pressure

a_l = thermal conductivity factor

ρ_v = steam density

t = time

R = vial radius

λ_l = heat conductivity factor

ΔT = a liquid overheat

k_0 = values of concentration, therefore

w_i = the diffusion velocity

10.1 INTRODUCTION

In present work the dynamics and heat and mass transfer of vapor bubble in a binary solution of liquids was studied for significant thermal, diffusion and inertial effect. Consider a two-temperature model of interphase heat exchange for the bubble liquid. This model assumes homogeneity of the temperature in phases. The intensity of heat transfer for one of the dispersed particles with an endless stream of carrier phase will be set by the dimensionless parameter of Nusselt Nu_l.

In this work dynamics and heat mass transfer of a steam bubbles in a binary mixture of liquid was studied. On the other hand simultaneously essential thermal, diffusion and ratchet effects were investigated. The dynamics and heat and mass transfer of vapor bubble in a binary solution of liquids, in [1], was studied for significant thermal, diffusion and inertial effect. It was assumed that binary mixture with a density ρ_l, consisting of components 1 and 2, respectively, the density ρ_1 and ρ_2. It was localized between limiting values for corresponding parameters of pure component. It showed that pressure differences and accordingly diffusion role were insignificant. In this work the influence of heat exchange and diffusion on weaken of this process were investigated.

10.2   MATERIALS AND METHODS

Bubble dynamics described by the Rayleigh equation [2]:

$$R\dot{w_l} + \frac{3}{2}w_l^2 = \frac{p_1 + p_2 - p_\infty - 2\sigma/R}{\rho_l} - 4v_1\frac{w_l}{R} \tag{1}$$

where p_1 and p_2—the pressure component of vapor in the bubble, p_∞—the pressure of the liquid away from the bubble, σ and v_1—surface tension coefficient of kinematic viscosity for the liquid.

Consider the condition of mass conservation at the interface. Mass flow j_i th component $(i = 1,2)$ of the interface $r = R(t)$ in j th phase per unit area and per unit of time and characterizes the intensity of the phase transition is given by:

$$j_i = \rho_i\left(\dot{R} - w_l - w_i\right), \ (i = 1,2) \tag{2}$$

where w_i—the diffusion velocity component on the surface of the bubble.

The relative motion of the components of the solution near the interface is determined by Fick's law:

$$\rho_1 w_1 = -\rho_2 w_2 = -\rho_l D\frac{\partial k}{\partial r}\bigg|_R \tag{3}$$

If we add equation (2), while considering that $\rho_1 + \rho_2 = \rho_l$ and draw the equation (3), we obtain

$$\dot{R} = w_l + \frac{j_1 + j_2}{\rho_l} \tag{4}$$

Multiplying the first equation (2) on ρ_2, the second in ρ_1 and subtract the second equation from the first. In view of (3) we obtain

$$k_R j_2 - (1 - k_R) j_1 = -\rho_l D \frac{\partial k}{\partial r}\bigg|_R$$

Here k_R—the concentration of the first component at the interface. [3-4] With the assumption of homogeneity of parameters inside the bubble changes in the mass of each component due to phase transformations can be written as

$$\frac{d}{dt}\left(\frac{4}{3}\pi R^3 \rho_i'\right) = 4\pi R^2 j_i$$

or

$$\frac{R}{3}\dot{\rho_i'} + R\dot{\rho_i'} = j_i, \ (i = 1, 2),$$ (5)

Express the composition of a binary mixture in mole fractions of the component relative to the total amount of substance in liquid phase

$$N = \frac{n_1}{n_1 + n_2}$$ (6)

The number of moles i th component n_i, which occupies the volume V, expressed in terms of its density

$$n_i = \frac{\rho_i V}{\mu_i}$$ (7)

Substituting (7) in (6), we obtain

$$N_1(k) = \frac{\mu_2 k}{\mu_2 k + \mu_1(1-k)}$$ (8)

By law, Raul partial pressure [5] of the component above the solution is proportional to its molar fraction in the liquid phase that is

$$p_1 = p_{S1}(T_v) N_1(k_R), \ p_2 = p_{S2}(T_v)\left[1 - N_1(k_R)\right]$$ (9)

Equations of state phases have the form:

$$p_i = BT_v \rho_i' / \mu_i, \ (i = 1, 2),$$ (10)

where B—gas constant, T_v—the temperature of steam, ρ_i'—the density of the mixture components in the vapor bubble μ_i—molecular weight, p_{Si}—saturation pressure. The boundary conditions $r = \infty$ and on a moving boundary can be written as:

$$k\big|_{r=\infty} = k_0, \ k\big|_{r=R} = k_R, \ T_l\big|_{r=\infty} = T_0, \ T_l\big|_{r=R} = T_v$$ (11)

$$j_1 l_1 + j_2 l_2 = \lambda_1 D \frac{\partial T_1}{\partial r}\bigg|_{r=R} \tag{12}$$

Where l_i—specific heat of vaporization.

By the definition of Nusselt parameter—the dimensionless parameter character-izing the ratio of particle size and the thickness of thermal boundary layer in the phase around the phase boundary are determined from additional considerations or from experience.

10.3   RESULTS AND DISCUSSION

The heat of the bubble's intensity with the flow of the carrier phase will be further specified as:

$$\left(\lambda_1 \frac{\partial T_1}{\partial r}\right)_{r=R} = Nu_1 \cdot \frac{\lambda_1 (T_0 - T_v)}{2R} \tag{13}$$

In [6-7] obtained an analytical expression for the Nusselt parameter:

$$Nu_1 = 2\sqrt{\frac{\omega R_0^2}{a_1}} = 2\sqrt{\frac{R_0}{a_1}}\sqrt{\frac{3\gamma p_0}{\rho_1}} = 2\sqrt{\sqrt{3\gamma} \cdot Pe_1}, \tag{14}$$

Where $a_1 = \frac{\lambda_1}{\rho_1 c_1}$—thermal diffusivity of fluid, $Pe_1 = \frac{R_0}{a_1}\sqrt{\frac{p_0}{\rho_1}}$—Peclet number.

The intensity of mass transfer of the bubble with the flow of the carrier phase will continue to ask by using the dimensionless parameter Sherwood SH:

$$\left(D \frac{\partial k}{\partial r}\right)_{r=R} = Sh \cdot \frac{D(k_0 - k_R)}{2R}$$

Here D—diffusion coefficient, k—the concentration of dissolved gas in liquid, the subscripts 0 and R refer to the parameters in an undisturbed state and at the interface. We define a parameter in the form of Sherwood [8]

$$Sh = 2\sqrt{\frac{\omega R_0^2}{D}} = 2\sqrt{\frac{R_0}{D}}\sqrt{\frac{3\gamma p_0}{\rho_1}} = 2\sqrt{\sqrt{3\gamma} \cdot Pe_D} \tag{15}$$

where $Pe_D = \frac{R_0}{D}\sqrt{\frac{p_0}{\rho_1}}$—diffusion Peclet number.

10.4   CONCLUSION

Pressure difference increased along with thermal dissipation and diffusion dissipation. Thus speed reduction and bubble growth considerably was high. It was higher than pure components of a mixture under the same conditions. The other condition was observed at growth and collapse of a steam bubble in water mixture of ethylene glycol. In this case the diffusion affected the resistance .It was led to breaking the speed of phase transformations. Growth and collapse rate of a bubble had much less than corresponding values (for pure components of a mixture). Structure and concentration of a component for binary solution are selected by practical consideration. The variation of structure happened when speed of phase transformations become low.

The systems of equations (1–15) are closed system of equations describing the dynamics and heat transfer of insoluble gas bubbles with liquid.

KEYWORDS

- **Binary mixture**
- **Fick's law**
- **Heat and mass transfer**
- **Homogeneity**
- **Nusselt parameter**

REFERENCES

1. Nagiyev, F. B. and Kadirov, B. A. (1986). B. A. Small oscillations of bubbles of two-component mixture in acoustic field. Bulletin Academy of Sciences Azerbaijan. Ser. Phys.Tech. Math. Sc. N1, 150-153.

2. Nagiyev, F. B. (1983). Linear theory of propagation of waves in binary bubbly mixture of liquids. Dep. in VINITI 17.01.86., N 405, 86, 120-128.

3. Nagiyev, F. B. (1989). The structure of stationary shock waves in boiling binary mixtures. Bulletin Academy of Sciences USSR. Mechanics of Liquid and Gas (MJG), N1 USSR, 81-87.

4. Nigmatulin, R. I. (1987). Dynamics of multiphase mediums. M. "Nauka", USSR 1, 2, pp. 67-78.

5. Nigmatulin, R. I., Nagiyev, F. B., and Khabeev, N. S. (1979). N. S. Vapor bubble collapse and amplification of shock waves in bubbly liquids. Proceedings "Gas and Wave Dynamics" N 3. Moscow State University, pp. 124-129.

6. Nigmatulin, R. I., Nagiyev, F. B., and Khabeev, N. S. (1980). Effective coefficients of heat transfer of bubbles radial pulsating in liquid. Thermo-mass-transfer two-phase systems. Proceedings of VI conf. Thermo-mass-transfer, vol. 5 Minsk, pp. 111-115.

7. Tikhonov, A. N. and Samarski, A. A. (1977). Equations of the mathematical physics. M, "Nauka", Moscow, USSR, p. 736.

8. Vargaftic, N. B. (1972). The directory of thermo physics properties of gases and liquids. M, "Nauka", USSR, pp. 67-79.

Index